Mechanical Behaviour of Engineering Materials

Mechanical Behaviour of Engineering Materials

Editor

Shrushti Omer

Mechanical Behaviour of Engineering Materials

Edited by **Shrushti Omer**

Printed in 2017

ISBN: 978-1-68117-224-8

Library of Congress Control Number: 2015936583

Contents

vi

Preface

How do engineering materials deform when bearing mechanical loads? To answer this crucial question, the book bridges the gap between continuum mechanics and materials science. The different kinds of material deformation (elasticity, plasticity, fracture, creep, fatigue) are explained in detail. The book also discusses the physical processes occurring during the deformation of all classes of engineering materials (metals, ceramics, polymers, and composites) and shows how these materials can be strengthened to meet the design requirements. It provides the knowledge needed in selecting the appropriate engineering material for a certain design problem. The reader will thus learn how to critically employ design rules and thus to avoid failure of mechanical components.

Editor

Mechanical Behaviour of Inconel 718 Thin-Walled Laser Welded Components for Aircraft Engines

Enrico Lertora, Chiara Mandolfino, and Carla Gambaro

Department of Mechanical Engineering, Polytechnic School of Genoa,
via all'Opera Pia 15, 16145 Genoa, Italy

ABSTRACT

Nickel alloys are very important in many aerospace applications, especially to manufacture gas turbines and aero engine components, where high strength and temperature resistance are necessary. These kinds of alloys have to be welded with high energy density processes, in order to preserve their high mechanical properties. In this work, CO_2 laser overlap joints between Inconel 718 sheets of limited thickness in the absence of postweld heat treatment were made. The main application of this kind of joint is the manufacturing of a helicopter engine component. In particular the aim was to obtain specific cross section geometry, necessary to overcome the mechanical stresses found in these working conditions without failure. Static and dynamic tests were performed to assess the welds and the parent material fatigue life behaviour. Furthermore, the life trend was identified. This research pointed out that a full joint shape control is possible by choosing proper welding parameters and that the laser beam process allows the maintenance of high tensile strength and ductility of Inconel 718 but caused many liquation microcracks in the heat affected zone (HAZ). In spite of these microcracks, the fatigue behaviour of the overlap welds complies with the technical specifications required by the application.

INTRODUCTION

Nickel alloys are generally used in highly aggressive environments, where qualities such as good strength and high temperature resistance become really important.

These kinds of alloys were developed in the early to mid-twentieth century, when the need to increase the efficiency of gas turbines, employed for civil and military applications, induced researchers to find new highly resistant alloys, which could offer higher strength levels, even in corrosive and high-temperature environments.

One of the most important materials employed for this kind of application is Inconel 718, a Ni-Cr-Fe based alloy characterized by excellent corrosion resistance combined with outstanding tensile, fatigue, and creep properties.

Unlike other complex Ni alloys, this material shows better welding characteristics, offering a higher resistance to postweld cracking [1, 2].

CO_2 laser beam welding (LBW) is one of the most widely used welding processes for Inconel 718.

The high density of energy allows the material to be melted rapidly: these results in narrow deep welds, high welding rates, and low heat input that minimize distortions and extension of the HAZ.

This welding process gives the possibility of realizing small thickness welds with a full joint shape control, even on different materials, and it can be easily included in a completely automated production system.

Thanks to its characteristics, LBW is witnessing wide development in many industrial sectors, in particular in the automotive [3, 4], aerospace [5], power-plant, and military ones, where it is mandatory to make really complex welds with high quality and production standards.

Some research work has focused on the implementation of Ni alloy laser welded joints, which mainly concern butt joints. The first scientific works carried out using the laser as a means to join the parts are those made by Gobbi et al. [6] and Cantello et al. [7] who compared the Nd-YAG and CO_2 laser techniques for the realization of high thickness Inconel 718 butt joints. During these research projects careful analysis of pre- and postwelding heat treatments was carried out to eliminate the metallurgical problems related to the formation of microcracks.

Ram et al. [8] analyzed the microstructure and mechanical properties of 2 mm thick Inconel 718 butt joints, welded using Nd-YAG pulsed laser, focusing on different post weld heat treatment.

Hong et al. [9] carried out a study of mechanical characteristics with tensile and fatigue tests conducted on 5 mm thick Inconel 718 butt joints. The joints were welded with different couples of power and welding speed and different postweld heat treatments, noting a considerable increase of the mechanical resistance of the joints treated after welding. The development of new Ni superalloys led Vishwakarma et al. [10] to study the heat treatment necessary for the elimination of metallurgical defects that develop during welding. These alloys were welded with the electron beam concentrated energy technique.

Several studies have been made concerning nickel and its weldability with concentrated energy. However, besides this great diffusion, information about the effect of the laser welding process on the fatigue properties of a high performance alloy, such as Inconel 718, is still difficult to find in the literature.

In the study above, an example of small thickness Inconel 718 LBW overlap joint, used for the manufacturing of an important helicopter engine component, is reported. This joint must be characterized by specific cross section geometry, necessary to overcome the mechanical stresses found in these working conditions without failure.

The limited thickness of the pieces to be joined does not allow postwelding heat treatments without altering the geometry of the piece; the strong distortions of the plates might seriously compromise the use of the component. In fact the heat treatment normally adopted provides solubilisation or aging at temperatures around 1000°C. In the literature there are no studies that take into account limited material thickness to be welded. Moreover, all studies concern only butt joints. Squillace et al. [11] studying the laser weldability of 1.6 mm plates Ti-6Al-4V alloys give an important contribution to understanding the parameters in order to obtain the correct macrogeometry of the joint.

The study of fatigue behaviour is still incomplete for butt joints and totally absent on Ni alloy lap joints. The innovation of this work lies mainly in the study of the fatigue behaviour of overlap laser joints between limited thickness Inconel 718 sheets in the absence of postweld heat treatment. Furthermore, the fatigue strength was also evaluated on the base material in order to be able to perform a comparison.

MATERIALS AND METHODS

The material used in this study is Inconel 718, an age-hardenable alloy characterized by a complex chemical composition, shown in Table 1 [12].

Table 1: Inconel 718 chemical composition [%] [12]

Ni	Cr	Co	Mo	Nb	Ti	Al	C	Mn	Si	B	Other	Fe
50.0 ÷ 55.0	17.0 ÷ 21.0	1.0	2.8 ÷ 3.3	4.75 ÷ 5.5	0.65 ÷ 1.15	0.20 ÷ 0.80	0.08	0.35	0.35	0.006	0.3 Cu	Bal.

Its high mechanical properties, guaranteed both at high and at low temperatures, are linked to the excess of Nb dissolved in the matrix, which results in the precipitation of the meta-stable body-centred tetragonal phase γ'' and of the stable orthorhombic phase Ni_3Nb (δ phase), following heat treatment [13].

In this study, as-rolled Inconel 718 sheets were lap-joined by a CO_2 laser welding process. Two different thicknesses of 0.4 mm and 1.6 mm were employed realizing dissimilar joints. Each plate was 300 mm in length and 210 mm in width.

The plates were welded using a 6 kW maximum power CO_2 ELEN-RTM machine, equipped with a nozzle delivering continuous argon shielding gas flow directly on the weld (Figure 1). Helium was used as plasma suppression gas.

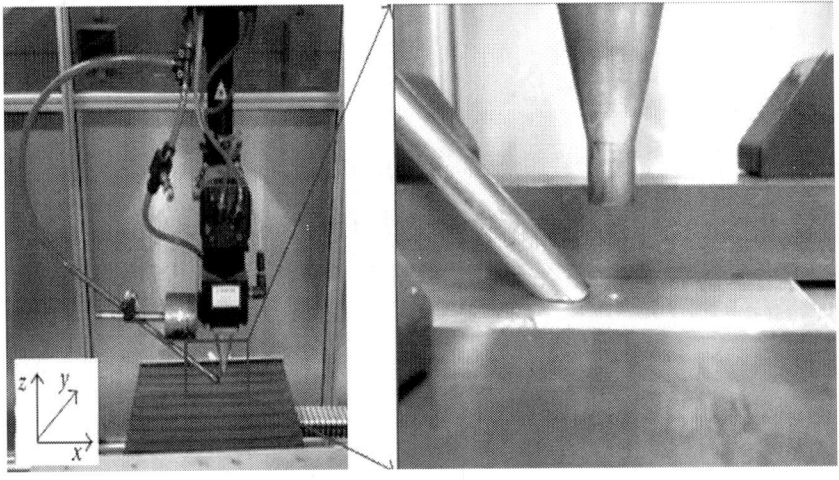

Figure 1: CO_2 ELEN-RTM machine and a detail of the shielding gas system.

A first work section was performed to create defect free welds, characterized by specific geometric parameters, required for the above mentioned industrial application. In particular, it was necessary to reach a weld width of 0.7-0.8 mm at the interface between the sheets and a weld penetration of 1.0–1.2 mm (Figure 2).

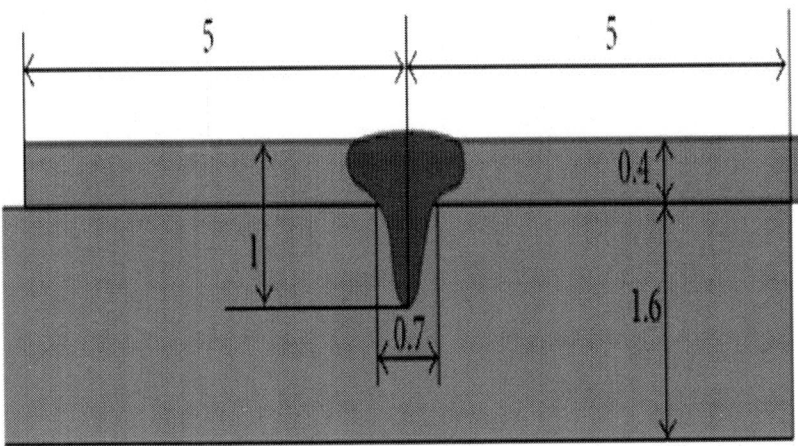

Figure 2: Weld cross section geometrical features.

To verify the achievement of this target, each weld was subjected to macrographic examination, performed with a Leica MZ6 modular stereomicroscope.

In the second part of the experimental test, mechanical properties of the specimens obtained from the welded sheets were investigated, with particular interest for the welds' fatigue behaviour. Parent material fatigue properties were also evaluated and compared to those for welds. In this case, Inconel 718 sheets, 1.6 mm thick, 40 mm wide, and between 190 and 200 mm long, were used. Static and dynamic tests were carried out on an Instron 8801 servo-hydraulic dynamic testing machine equipped with a 50 kN load cell.

The weld static and fatigue behaviours were evaluated by testing rectangular cross section specimens, 40 mm wide, obtained by cutting the welded plates with a shearing machine (Figure 3). The load-bearing cross section was 28 mm².

(a)

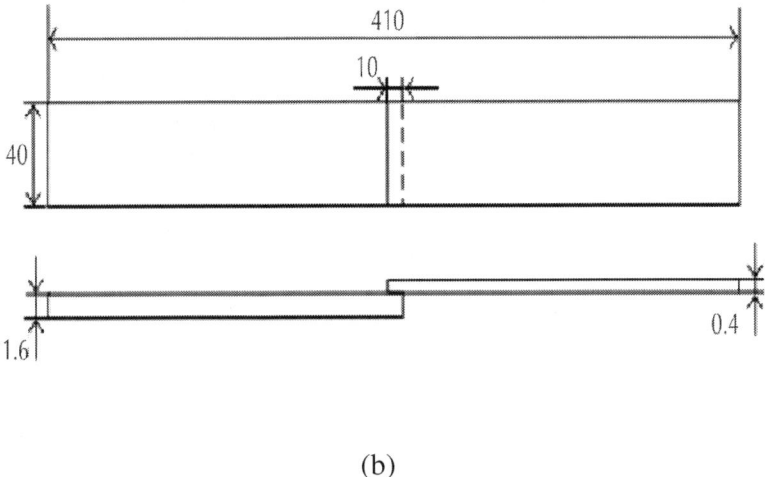

(b)

Figure 3: Geometric features of the fatigue specimen.

The specimens were inserted in the tensile machine grips by using two metallic tabs of 0.4 and 1.6 mm. In order to avoid vibration of the thinner sheet at stress frequency, a longer metallic tab was used (Figure 4).

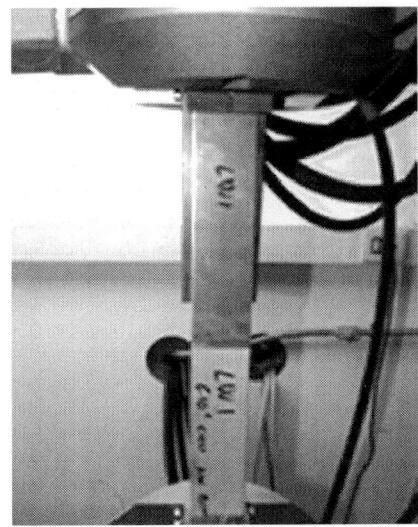

(a)

(b)

(c)

Figure 4: Specimen inserted in the machine grips and metallic tabs used.

In this experimental session, a total of 33 test specimens (21 obtained from the welds and 11 from the parent material) were subjected to different sinusoidal stress cycles, characterized by a load ratio $\mu = 0.1$ and a frequency f of 10 Hz.

The macro- and microstructures of the tested specimens were examined after the fatigue experimental section, by using a Leica MeF3 microscope.

All samples were mechanically polished using a diamond compound up to 1 μm grain size and then etched with a solution composed of 50 cc H_2O, 50 cc HCl, and 1.2 cc HNO_3, heated to a temperature of between 70° and 85°C.

RESULTS AND DISCUSSION

Characteristics of CO_2 Laser Welding

The welding session highlighted some important correlations between the welding parameters (laser power and welding speed) and the geometrical joint aspects (welding penetration and weld width), in accordance with other experimental activities [4, 14, 15]. Looking at Figure 5(a), it is possible to notice that, for a constant value of welding speed, the joint dimensions (weld penetration and width) are directly proportional to the laser power input. Figure 5(b), instead, shows that, for a constant laser power input, the main dimensions of the joint are in inverse proportion to the welding speed.

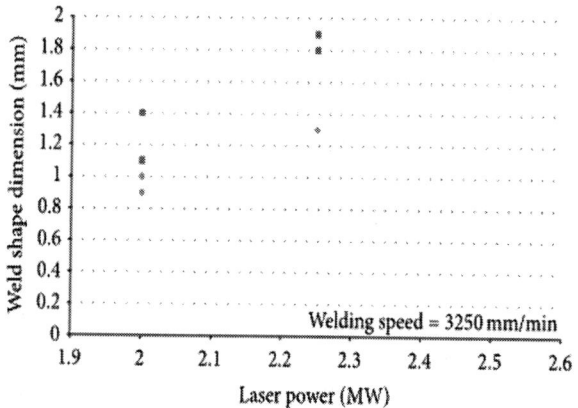

(a)

(b)

Figure 5: Penetration and width of welds in function of laser power (a) and of welding speed (b).

From this welding campaign the optimal parameters to create the target geometry were chosen. In Table 2 the optimal joint shape was reported as well as the welding parameters used to create it.

Table 2: Optimal welding parameters and resulting welded specimen

Welding parameters			
Welding speed [mm/min]	4000		
Power [W]	2500		
Primary gas (He) [l/h]	1500		
Shielding gas (Ar) [l/h]	15		
Focus [mm]	0.2		
Weld macrography			
	Dimensional specifications	Welded specimen	Target range
	Welding penetration [mm]	1.2	1.0 ÷ 1.2
	Weld width at 0.4 mm [mm]	0.7	0.7 ÷ 0.8
		No defects observed	

Mechanical Properties and Fatigue Life Evaluation

The first part of the experimental session focused on the static characterization of the lap joints.

The tensile properties of the welds were assessed by testing three specimens obtained from the welded plates. The samples had the same geometry as the fatigue specimens, shown in Figure 3. The results of the tensile tests are reported in Figure 6. All the samples failed near the weld line on the thinnest sheet side.

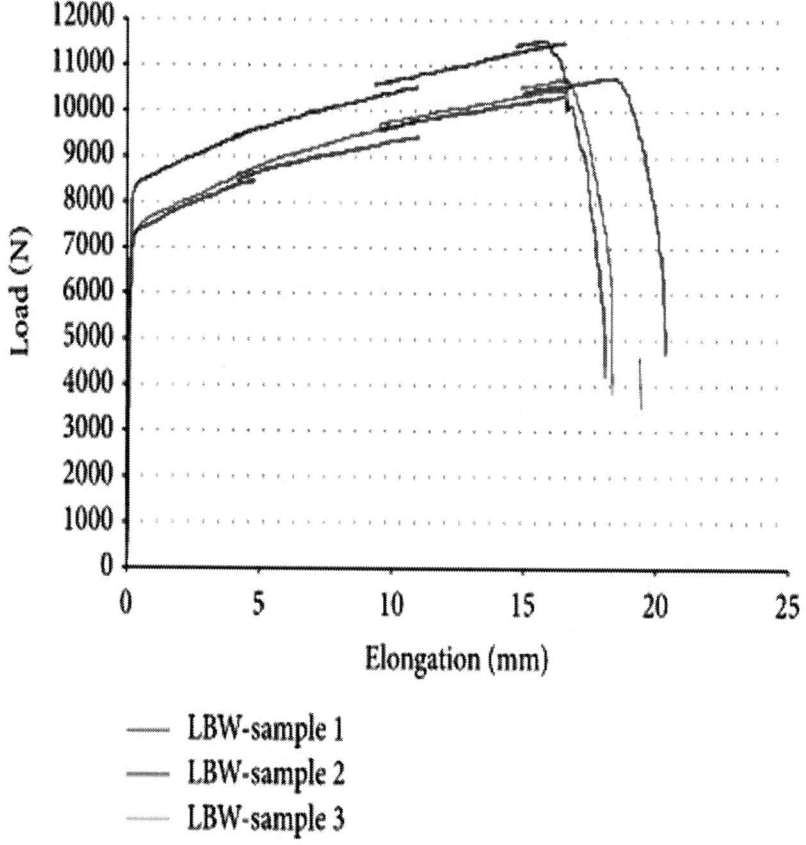

Figure 6: LBW tensile behaviour.

As can be seen, the joints present good ductility, though it is obviously lower than the parent material [12]. The three samples showed good tensile strength, respectively, of 382.24, 380.27, and 412.44 MPa, with a resulting average value of 391.65 MPa.

Then dynamic tensile tests were carried out to evaluate the behaviour of laser beam welds and also the parent material in terms of cyclic strength performance.

The data collected were processed and reported in a diagram (Figure 7), indicating the number of cycles to failure (N) in function of the stress amplitude (σ_a), allowing us to point out the fatigue life trend of the welds and of the Inconel 718.

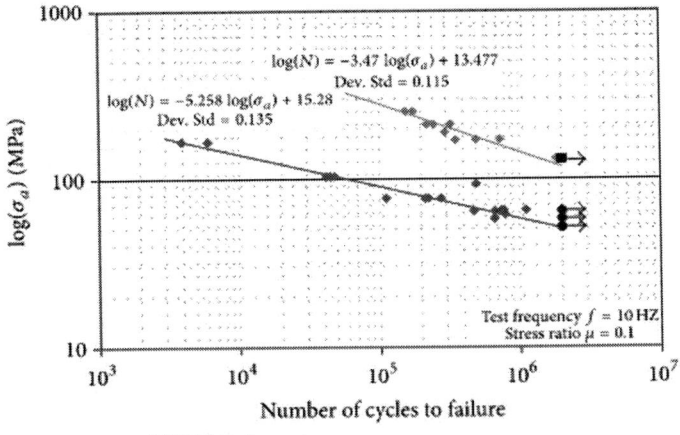

Figure 7: Fatigue behaviour of LBW and of Inconel 718.

In order to obtain a good demonstration of the joint fatigue behaviour, seven different stress amplitude levels were used, while only 5 stress levels were considered for the parent material.

In the case of the LBW, the first load step investigated (σ_a = 164.83 MPa) resulted in an early fracture of the three test samples, which failed after an average of only 4590 cycles. Hence, lower step levels were tested, in order to achieve a fatigue endurance of about $2 \cdot 10^6$ cycles, which was assumed to be the value representing an unlimited life of the joint.

A statistical study of the fatigue data was performed to define a linearized stress-life curve, as described by the standard practice reported in ASTM E 739-91 [16]. This method allowed definition of two curves—identified as "P50^I" and "P50^{II}"—respectively, for the welds and for the parent material, corresponding to a survival probability of 50% for the specimens.

Analysis of these curves allowed us to define the fatigue limit of the LBW and of the Inconel 718, which turned out to be, respectively, 51 and 116.9 MPa.

From an analysis of the diagram, it is possible to notice a higher dispersion of the weld data compared to that of the parent material. This aspect is also confirmed by the higher standard deviation value of the laser beam specimens.

All the tested specimens failed in the same way: the crack started from the 0.4 mm Inconel sheet side, at the interface between the fused zone (FZ) and the base material (BM), which is a very narrow HAZ and developed along the weld line or in the base material, until the specimen was completely broken (Figure 8).

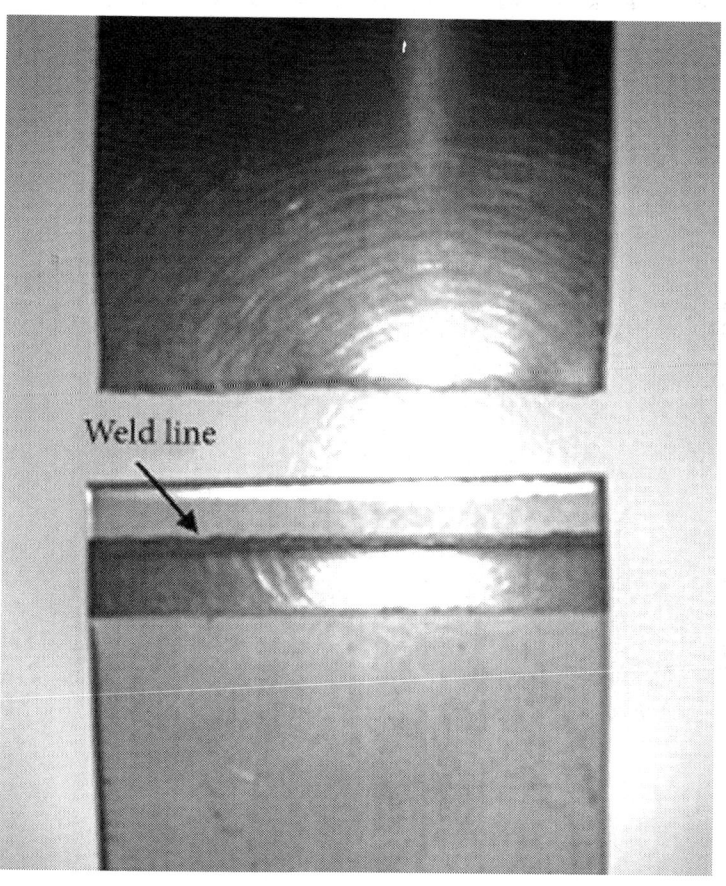

(a)

$$G\varepsilon(t) + \mu\frac{d\varepsilon(t)}{dt} = \sigma(t)$$

(3)

In formula (3), themeaning of (t), (t), G, and μ was the same as the above marked.

Analysis on Stress Relaxation and Creepage of Maxwell Model and Kelvin Model

Analysis onMaxwellModel. In the case of creepage, (t) = σ_0, formula (2) turned to

$$\frac{d\varepsilon(t)}{dt} = \frac{\sigma_0}{\mu}$$

(4)

In the case of stress relaxation, $d\varepsilon(t)/dt = 0$, formula (2) turned to

$$\frac{1}{G}\frac{d\sigma(t)}{dt} + \frac{\sigma(t)}{\mu} = 0$$

(5)

Obtaining the equation of stress changed with time after integration,

$$\sigma(t) = ae^{-(1/b)t} + c$$

(6)

Among it, a, b, and c are constants, and $b = \mu/G$, $c = 0$.

As was shown in formula (4), strain changed uniformly with time in Maxwell model in the case of creepage, which did not match with the actual situation of the viscoelastic material. Formula (6) suggested that the stress decayed exponentially with time, but it did not meet the general situation of viscoelastic material, as the creepage behavior of the material was very complicated. Furthermore, the actual stress relaxation behavior of viscoelastic material could not decay to zero for a long time. In a word, a simple exponential decayed item could not represent the real situation of creepage behavior.

Analysis on Kelvin Model. In the case of stress relaxation, formula (3) turned to

$$G\varepsilon\left(t\right) = \sigma\left(t\right)$$

(7)

It only showed the mechanical behavior of spring element, fully met with Hooke's law, which did not accord with the actual situation of viscoelastic material, which also suggested that the situation of stress relaxation was not suitable for this formula.

In the case of creepage, $\sigma\left(t\right) = \sigma_0$, formula (3) turned to

$$\frac{d\varepsilon\left(t\right)}{dt} = \frac{\sigma_0}{\mu} - \frac{G}{\mu}\varepsilon\left(t\right)$$

(8)

After integration of both sides of the above formula, obtain the equation in the following form:

$$\varepsilon\left(t\right) = de^{-1/f^t} + g$$

(9)

Among it, $d = g = \sigma_0/G$.

As the initial deformation is zero, it could be obtained:

$$\varepsilon\left(t\right) = d\left(1 - e^{-1/f^t}\right)$$

(10)

Among it, the meaning of the parameters was the same as the above.

As was shown in formulas (7) and (8), Kelvin model could better describe the creepage property of viscoelastic material but could not exactly describe the relaxation property of viscoelastic materials.

Constructing a Custom Composite Model

Composite model referred to be connected by multiple Maxwell models, Kelvin model, and spring in parallel, the study was showed, composite model integrated the advantages of the above two models, and in accordance with the following constitutive relation formula [15],

$$\sigma\left(t\right) + \sum_{m=1}^{M} a_m \frac{d^m \sigma\left(t\right)}{dt^m} = G\varepsilon\left(t\right) + \sum_{n=1}^{N} b_n \frac{d^n \varepsilon\left(t\right)}{dt^n}$$

(11)

(11)

At, $M = 2$, $N = 2$ (two Maxwell models and a spring connected in parallel), the above formula could be expressed as

$$\sigma(t) + a_1 \frac{d\sigma(t)}{dt} + a_2 \frac{d^2\sigma(t)}{dt^2} = G\varepsilon(t) + b_1 \frac{d\varepsilon(t)}{dt} + b_2 \frac{d^2\varepsilon(t)}{dt^2} \qquad (12)$$

As could be seen, along with the number of Maxwell models and springs in parallel, more unknown quantities were introduced and required more groups of data; solving was also becoming very difficult. So this model was constructed by selection of formula (12), as was shown in Figure 2.

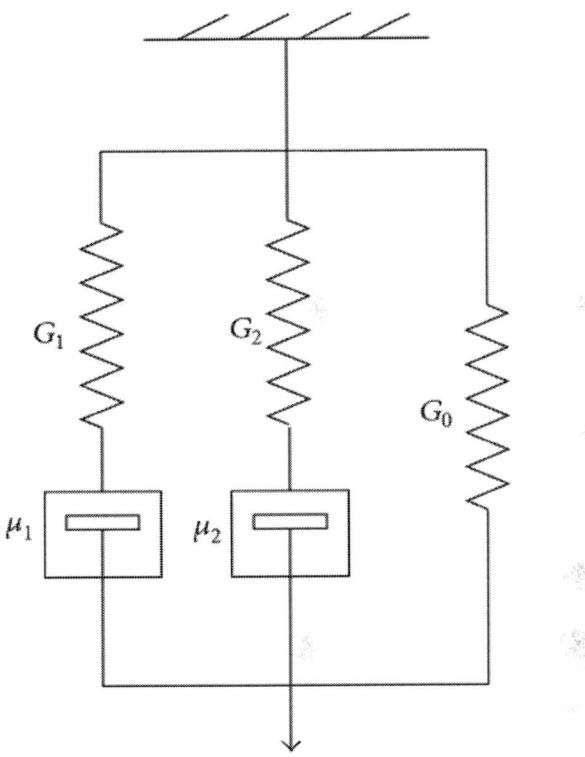

Figure 2: Constructed composite model.

In the case of creepage ($d\varepsilon(t)/dt = 0$, $\varepsilon = \varepsilon_0$), it could be concluded that

$$\sigma\left(t\right) + a_1 \frac{d\sigma\left(t\right)}{dt} + a_2 \frac{d^2\sigma\left(t\right)}{dt^2} = G\varepsilon_0$$

(13)

In the case of stress relaxation ($\sigma\left(t\right) = \sigma_0$), it could be concluded that

$$\sigma_0 = G\varepsilon\left(t\right) + b_1 \frac{d\varepsilon\left(t\right)}{dt} + b_2 \frac{|d^2\varepsilon\left(t\right)}{dt^2}$$

(14)

Model Validation

According to Maxwell model, fitting only needs to obtain the data of fixed strain, the fitting results were as shown in Figure 3, and the formula of function fitting relationship was as follows:

$$\sigma\left(t\right) = ae^{(1/-b)t} + c = 0.45418e^{(1/-22.66264)t}$$

(15)

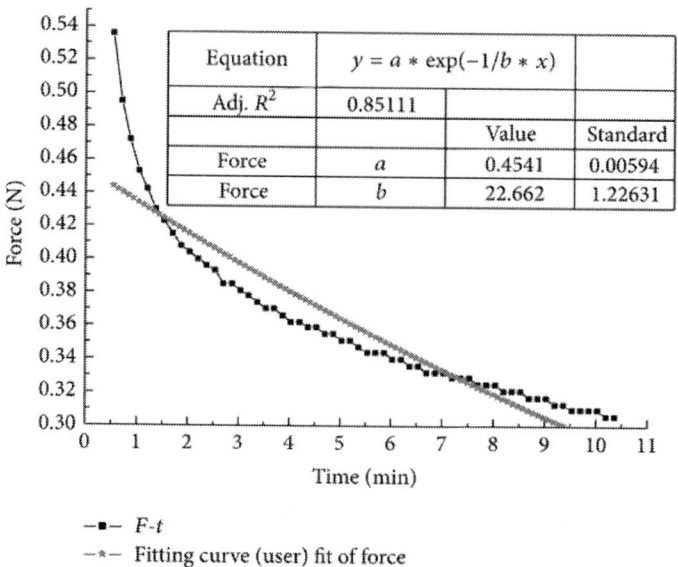

Equation	$y = a * \exp(-1/b * x)$		
Adj. R^2	0.85111		
		Value	Standard
Force	a	0.4541	0.00594
Force	b	22.662	1.22631

—■— F-t
—*— Fitting curve (user) fit of force

Figure 3: *F-t* fitting curve of Maxwell model.

Substituting $\sigma\left(t\right)$ into formula (4), the relationship formulaof $\left(t\right)$ was obtained. The model ratio could only obtain the ratio of μ and G, which was 22.66264, and it could not solve out the specific numerical values of μ and G.

According to the Kelvin model, fitting only used the data obtained from fixed stress, the fitting result was as shown in Figure 4, and the formula of function relationship obtained by origin fitting was

$$\varepsilon(t) = ae^{(1/-b)t} + c = -2.79136e^{(1/-4.52342)t} - 27.0611 \quad (16)$$

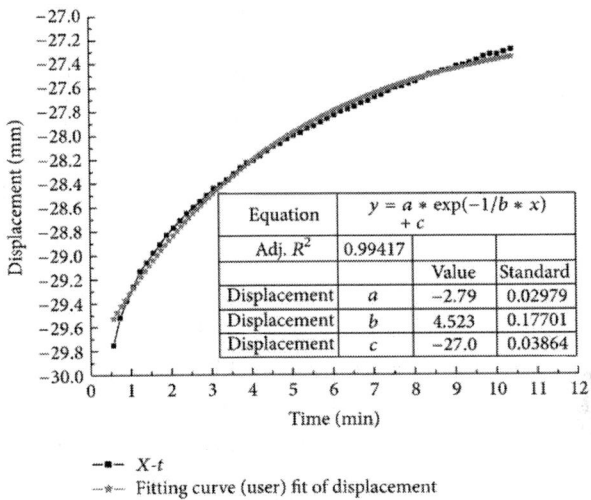

Equation	$y = a * \exp(-1/b * x) + c$	
Adj. R^2	0.99417	
	Value	Standard
Displacement a	-2.79	0.02979
Displacement b	4.523	0.17701
Displacement c	-27.0	0.03864

— X-t
—*— Fitting curve (user) fit of displacement

Figure 4: X-t fitting curve of Kelvin model.

After solving,

$$G = 0.01667, \qquad \mu = 0.07539 \quad (17)$$

The fitting curves of the constructed model were shown as in Figures 5 and 6, obtaining the following formula:

$$\sigma(t) = 0.32375e^{(1/-0.44747)t} + 0.17208e^{(1/-6.07104)t} - 0.27647,$$

$$\varepsilon(t) = -1.55703e^{(1/-0.44747)t} - 2.72363e^{(1/-6.07104)t}$$

$$- 26.81475. \quad (18)$$

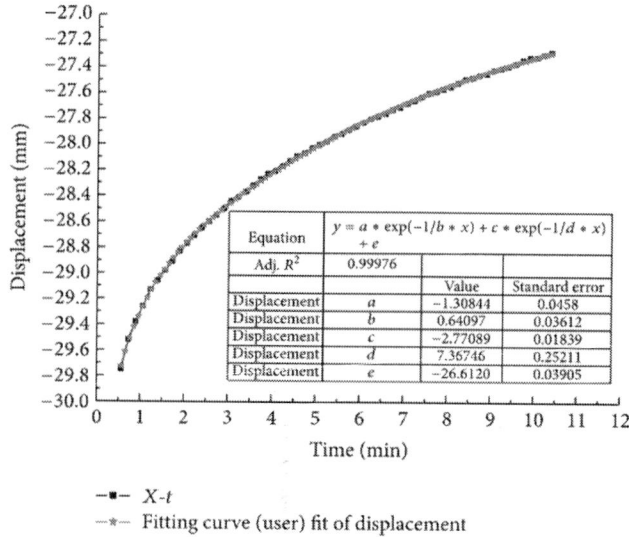

Figure 5: *X-t* fitting curve of constructed composite model.

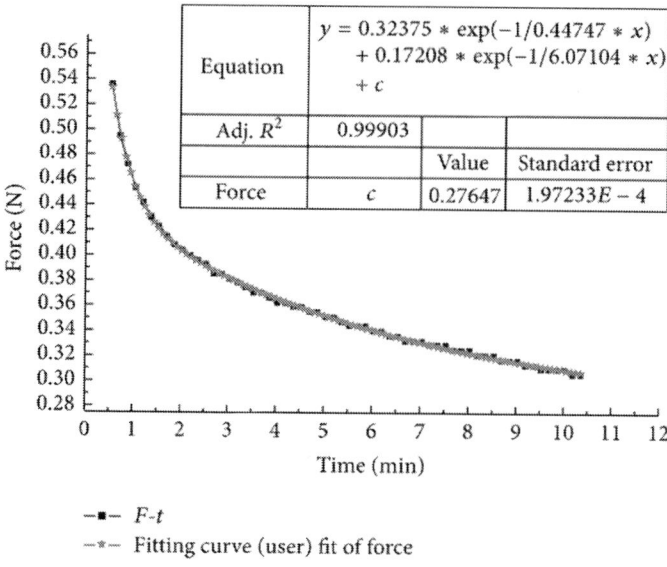

Figure 6: *F-t* fitting curve of constructed model.

After solving,

$$a_1 = 0.84793, \qquad a_2 = 7.5463 \times 10^{-4},$$

$$b_1 = 0.10864, \qquad b_2 = 7.54614 \times 10^{-4},$$

$$G = 0.01031. \tag{19}$$

The value of R represented the fitting effect; through the comparison of R values, it was calculated, relative to the one-sided description of Maxwell model and Kelvin model of viscoelastic food material; the fitting result of constructed model was the best, while the Maxwell model was the worst; the next was Kelvin model.

In the different compression ratio (5%, 10%, and 15%), fitting cases of Maxwell model and constructed model were further studied, as shown in Figures 7 and 8; the fitting correlation coefficients in the compression ratio of 5%, 10%, and 15% of Maxwell model were 0.85813, 0.85111, and 0.81914; the fitting effect was getting more and more worse with the increasing deformation; the fitting correlation coefficients in the compression ratio of 5%, 10%, and 15% of constructed model were 0.99714, 0.99896, and 0.99895; in the case of increasing deformation, the value of R was close to 1 and the fitting effect was better.

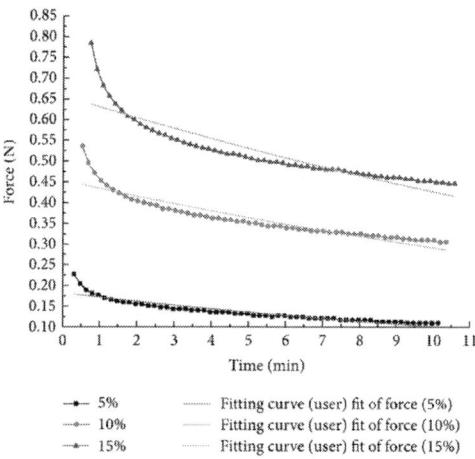

Figure 7: Fitting curves of Maxwell model in different compression ratio.

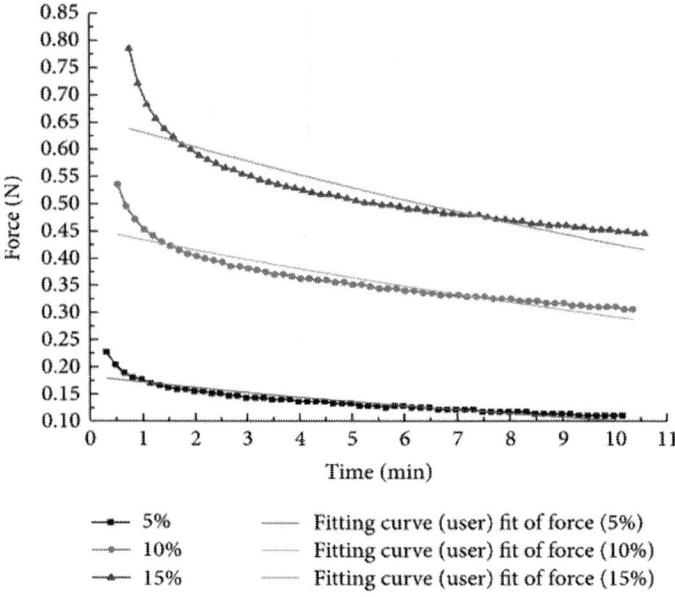

Figure 8: Fitting curves of constructed model in different compression ratio.

CONCLUSIONS

In this paper, through the stress-strain relaxation experiments of French roll produced by Fujian Dali Food Group Company, separately by Maxwell model, Kelvin model, and custom composite model, we studied mechanical behavior in the mode of compressed chewing, and established the stress-strain model in the mode of compressed chewing.

- In explaining the mechanical behavior of the viscoelastic food material under large deformation, the accuracy of Maxwell model and Kelvin model was not high.

- The constructed custom model through the stress relaxation experiments and strain relaxation experiments, compared to Maxwell model and Kelvin model, had a higher accuracy, which was better able to explain the mechanical behavior of this bread in the mode of chewing with large deformation.

- The method of constructed composite model was also applicable to construct the stress-strain model of other viscoelastic materials under different modes of masticatory, providing a more accurate theoretical model for the formulating of processing technology and selection of processing parameters of viscoelastic material.

ACKNOWLEDGMENTS

This study is supported by Jiangsu Cooperative Innovation Project (Grant no. SBY201320360) and Jiangsu Postdoctoral Fund (Grant no. 1201041c).

REFERENCES

1. X. Q. Zhou and J. R. Yi, "Effect of, different fermentation conditions on texture and rheological properties of fermented bean curd," Science and Technology of Food Industry, vol. 35, no. 7, pp. 217–220, 2014.

2. L. Liang, Y. P. Li, and Y. M. Guo, "Study on wheat stalk, experimental study on the viscoelastic properties," Agricultural Mechanization Research, vol. 33, no. 5, pp. 174–177, 2011.

3. L. Ma, J. Wu, D. Q. Zhao, et al., "Beef in Brown Sauce stress relaxation experiment," Food and Fermentation Industry, vol. 39, no. 1, pp. 221–224, 2013.

4. X. W. Zhao, Y. M. Wei, and S. K. Du, "Review on mathematical models of food property changes induced by extrusion," Transactions of the Chinese Society of Agricultural Engineering, vol. 24, no. 10, pp. 301–307, 2008. · ·

5. M. Shiozawa, H. Taniguchi, H. Hayashi, et al., "Differences in chewing behavior during mastication of foods with different textures," Journal of Texture Studies, vol. 44, no. 1, pp. 45–55, 2013. · ·

6. R.-T. Zhan, Z.-X. Li, and L. Wang, "A fractional differential constitutive model for dynamic stress intensity factors of an anti-plane crack in viscoelastic materials," Acta Mechanica Sinica, vol. 30, no. 3, pp. 403–409, 2014. · · View at MathSciNet

7.   H. Watanabe, T. Sato, K. Osaki et al., "Rheological images of poly(vinyl chloride) gels. 4. Nonlinear behavior in a critical gel state," Macromolecules, vol. 31, no. 13, pp. 4198–4204, 1998. · ·

8.   V. H. Rolón-Garrido and M. H. Wagner, "The damping function in rheology," Rheologica Acta, vol. 48, no. 3, pp. 245–284, 2009. · ·

9.   F. Renaud, J.-L. Dion, G. Chevallier, I. Tawfiq, and R. Lemaire, "A new identification method of viscoelastic behavior: application to the generalized Maxwell model," Mechanical Systems and Signal Processing, vol. 25, no. 3, pp. 991–1010, 2011. · ·

10.   C. J. Shuai, J. A. Duan, and J. A. Wang, "Method of establishing generalized Maxwell model for viscoelastic material," Chinese Journal of Theoretical and Applied Mechanics, vol. 38, no. 4, pp. 565–569, 2006.

11.   J. Ren and P. Guo, "A new mathematical model for pressure transient analysis in stress-sensitive reservoirs," Mathematical Problems in Engineering, vol. 2014, Article ID 485028, 14 pages, 2014. · ·

12.   J. Yu, D. Ma, and H. Lu, "Analysis on nonlinear stress-growth data for shear flow of starch material with shear process," Mathematical Problems in Engineering, vol. 2013, Article ID 406018, 5 pages, 2013. · ·

13.   C.-Y. Lee and M.-I. Kuo, "Effect of -polyglutamate on the rheological properties and microstructure of tofu," Food Hydrocolloids, vol. 25, no. 5, pp. 1034–1040, 2011. ·

14.   W. Wang, Z. Lu, Y. Cao, J. Chen, J. Wang, and Q. Zheng, "Investigation and prediction on the nonlinear viscoelastic behaviors of nylon1212 toughened with elastomer," Journal of Applied Polymer Science, vol. 123, no. 3, pp. 1283–1292, 2012. · ·

15.   I. M. Ward, Mechanical Properties of Solid Polymers, Science Press, 1980.

16.   X. Li, Fractional Oder Integral Models and Numerical Methods of Several Kinds of Viscoelastic Material, Ningxia University, 2013.

17.   C. Chung, K. Olson, B. Degner, and D. J. McClements, "Textural properties of model food sauces: correlation between simulated

mastication and sensory evaluation methods," Food Research International, vol. 51, no. 1, pp. 310–320, 2013. · ·

18. Y. P. Chu, "Research on Application of texture analyzer in food quality evaluation," Cereal and Feed Industry, no. 7, pp. 40–42, 2003.

Chapter 3

Effect of Aerogel Particle Concentration on Mechanical Behavior of Impregnated RTV 655 Compound Material for Aerospace Applications

Firouzeh Sabri[1] Jeffrey G. Marchetta[2] K. M. Rifat Faysal1 Andrew Brock[1], and Esra Roan[2]

[1]Department of Physics, University of Memphis, Memphis, TN 38152, USA

[2]Department of Mechanical Engineering, University of Memphis, Memphis, TN 38152, USA

ABSTRACT

Aerogels are a unique class of materials with superior thermal and mechanical properties particularly suitable for insulating and cryogenic storage applications. It is possible to overcome geometrical restrictions imposed by the rigidity of monolithic polyurea cross-linked silica aerogels by encapsulating micrometer-sized particles in a chemically resistant thermally insulating elastomeric "sleeve." The ultimate limiting factor for the compound material's performance is the effect of aerogel particles on the mechanical behavior of the compound material which needs to be fully characterized. The effect of size and concentration of aerogel microparticles on the tensile behavior of aerogel impregnated RTV655 samples was explored both at room temperature and at 77 K. Aerogel microparticles were created using a step-pulse pulverizing technique resulting in particle diameters between 425 µm and 90 µm and subsequently embedded in an RTV 655 elastomeric matrix. Aerogel particle concentrations of 25, 50, and 75 wt% were subjected to tensile tests and behavior of the compound material was investigated. Room temperature and cryogenic temperature studies revealed a compound material with rupture load values dependent on (1) microparticle size and (2) microparticle concentration. Results presented show how the stress elongation behavior depends on each parameter.

INTRODUCTION

Long-distance space travel is currently limited due to availability of fuel for such missions. Current technology relies on storage of cryogenic fuel in metallic containers, not practical for large quantities necessary for long-term travel. Polymeric materials are a highly adaptable group of materials with a broad range of applications that span the biomedical field to the space industry [1–3]. The versatility of polymer synthesis and processing has provided an excellent platform for the development of custom-designed, application-specific polymers with unique properties. Most polymers however suffer from low mechanical stiffness [4–6] and have limited load bearing capabilities which in turn limits their use but has been addressed in some studies by incorporating micro- and nanosized particles into the polymeric matrix [4,7–10]. The final mechanical strength of the impregnated polymers strongly depends on particle parameters such as particle size, geometry, concentration [11–14], and stress transfer between the encapsulated particles and the polymer matrix [15]. In general, if the applied stress is transferred effectively from the matrix to the particles the result will be a stronger material [16–20]. Since the introduction of micronanoparticles into a polymer matrix can have a variety of effects on the overall behavior of the material, each particle-polymer combination must be independently and thoroughly characterized and the limits of the material tested under different environmental conditions.

Aerogels are a relatively new class of materials with unique material properties particularly suitable for the aerospace industry. Aerogels are currently the best known solid thermal insulators and have been explored extensively for insulation applications both in the monolith form and in particle and bead geometries [21–23]. Low concentration (<15%) native aerogel-impregnated polymers have been investigated previously for thermal insulation applications [24, 25] and demonstrated great promise for applications such as cryogenic tank systems and insulation on the Mars exploration rovers [26]. Among the various types of aerogels, polyurea cross-linked silica aerogel (PCSA) is an attractive candidate for this study due to its relatively lower production cost, compatibility with green technology, and most importantly enhanced mechanical strength accomplished by covalent cross-linking of the polymer chains while retaining a high degree of porosity [27, 28]. The PCSA is also hydrophobic in comparison to the native silica aerogels making this type of aerogel most suitable for applications where there is a possibility for aerogel components to come into contact with liquids.

Silicone rubbers on the other hand are known to exhibit a high resistance and tolerance to UV radiation, excellent chemical stability, good thermal insulation characteristics, and a high tensile and tearing strength over a wide range of temperatures [29–31]. The room temperature vulcanizing types of elastomers are easy to manufacture and process in large quantities with a wide variety of surface and bulk properties to choose from. Additionally, polydimethylsiloxane-(PDMS-) based polymers are still one of the most suitable elastomers for cryogenic applications due to their low glass transition temperatures [32]. By combining PCSA and RTV 655, two space-qualified materials used previously for aerospace missions [33–39], it is possible to create compound materials with unique and tunable physical and thermal properties applicable to future space exploration as well as terrestrial applications requiring expandable lightweight insulating material.

In this study, the authors fully characterized the effect of high loading and impregnation levels of micron-sized particles on the mechanical behavior of RTV 655 elastomer at two critical temperatures: room temperature and 77 K. Microparticle impregnation levels (IL) of 25, 50, and 75 wt% were prepared, tested, and compared with neat RTV 655. The aerogel microparticles were created using a step-pulse pulverizing technique appropriate for creating large quantities of particles. Aerogel particles were separated mechanically into three size ranges: $d_1 = 300–425 \mu m$, $d_2 = 180–300 \mu m$, and $d_3 = 90–180 \mu m$. The tensile behavior of the compound material was investigated as a function of (a) aerogel particle concentration and (b) aerogel particle size. This study shows that for low IL the aerogel particle size dominates the tensile behavior of the compound material while at high IL the behavior is dominated by the concentration of the particles rather than their physical size.

MATERIALS AND METHODS

Aerogel Synthesis and Microparticle Formation

Polyurea cross-linked silica aerogels were synthesized and molded in 3 mL syringes as described in detail previously [32] and dried supercritically in a Quorum Technology's E3100 critical point drier. The bulk density of aerogel cylinders was determined by volumetric and gravimetric measurements and fell in the range of 0.317–0.50 gm/cm^3. The translucency (in this case is related to degree of porosity) of each aerogel batch was evaluated by means of a Beckman DU-60 UV-Vis spectrophotometer under atmospheric conditions. A total of nine batches were synthesized for this study, each batch producing ten 3 mL aerogel cylinders. The density of each monolithic sample was calculated before the pulverization step.

Synthesized aerogel cylinders were next pulverized in an IKA Works Inc. grinder model A11B1S1 with an IKA A11.1 blade for a total of 60 s in 30 s intervals to avoid internal heating. The optimum pulverization time was established first on a test batch of aerogels. The aerogel powder was then passed through ASM E-11 sieves (Fisher Scientific Inc.) with mesh sizes 425 μm, 300 μm, 180 μm, 90 μm, 45 μm, and 20 μm. It was observed that over 60% of the aerogel powder mass that was created was evenly distributed in the ranges of $d_1 = 300–425 \mu m$, $d_2 = 180–300 \mu m$, and $d_3 = 90–180 \mu m$ and only 20% of the total mass was distributed in the size ranges below 90 μm therefore limiting the PCSA particle sizes for the study to ranges greater than 90 μm. Microparticles were also imaged with a scanning electron microscope periodically. The percentage of mass distribution for each pulverized batch is shown in Figure 1.

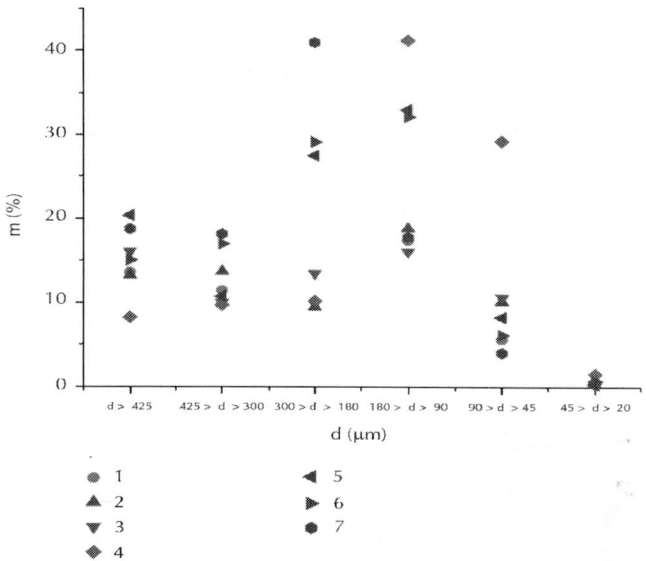

Figure 1: Percent mass distribution of PCSA particles for different diameter ranges for seven separate synthesis batches, after pulverization stage (t=60sec). The majority of PCSA microparticles created by step-pulse pulverization technique were greater than 90 μm in diameter.

Aerogel-Impregnated RTV 655 Sample Preparation

First, a ratio of 10:1 RTV 655 prepolymer to cross-linker was mixed thoroughly as instructed by the manufacturer and completely outgassed in a Blue-M Precision vacuum oven. Next, equal parts of the silicone mix were poured into three separate containers and precalculated amounts of PCSA particles of three different size ranges (d_1 = 300–425 μm, d_2 = 180–300 μm, and d_3 = 90–180 μm) were introduced into the silicone mix such that for each particle size range samples with a 25%, 50%, and 75% (weight percent of PCSA particles to RTV 655) were prepared. The 75% concentration was the maximum concentration that would allow full encapsulation of the microparticles and complete curing of the compound material. The neat (0% PCSA impregnation level) and impregnated RTV 655 slurries were thoroughly mixed, outgassed, and poured into SPI-A2 polished (The Mold Polishing Company, NJ), ASTM D1708 custom-designed dog-bone molds shown in Figure 2. Great care was taken during the mixing process to avoid aerogel cluster formations. The formation of agglomerates was prevented as much as possible by microsieving the microparticles before adding them to the RTV 655 and careful and consistent mixing upon adding to the RTV 655 with mechanical agitation during the mixing phase in order to prevent (as much as possible) agglomerates from forming. The RTV 655+PCSA mixtures were outgassed for the final time at room temperature and then cured at 100°C for 1 hr. Molds were then removed from the vacuum oven and allowed to cool down prior to removal of samples from molds. Table 1 summarizes all sample groups synthesized and prepared for this study.

Table 1: Summary of sample types prepared for this study

Sample group	Number of batches	Ingredients	PCSA diameter d=(μm)	% PCSA concentration by wt.
1	3	RTV 655		0%
		RTV 655 & PCSA	425–300	25%
				50%
				75%
2	3	RTV 655		0%
		RTV 655 & PCSA	300–180	25%
				50%
				75%
3	3	RTV 655		0%
		RTV 655 & PCSA	180–90	25%
				50%
				75%

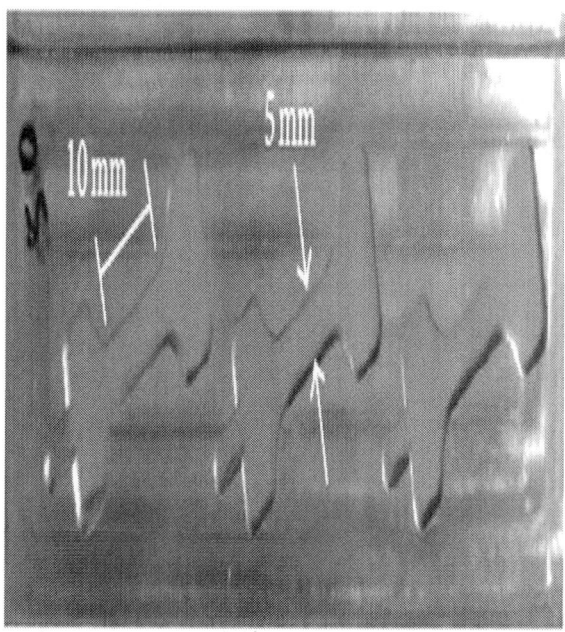

Figure 2: Highly polished SPI-A2 aluminum molds for curing of neat and impregnated RTV 655 samples for tensile testing. Dog-bones were cut according to ASTM D1708 geometrical specifications.

Room Temperature and Low Temperature Tensile Measurements

The tensile behavior of the neat and PCSA-impregnated RTV 655 samples was evaluated by means of an ESM 301 (Mark-10 Inc.) bench top tensile tester both at room temperature and at 77 K, according to the D1708-06a ASTM standard for testing of polymers. Data acquisition and manipulation were performed using the ESM 301 manipulation software Measure. The instrument was fully calibrated following manufacturer's recommendations prior to sample testing. For low temperature measurements a custom-designed cryogenic chamber was used. The sample under investigation was positioned between custom-designed stainless steel clamps and the assembly was positioned inside the chamber. Liquid nitrogen was then poured inside the chamber and once thermal equilibrium was accomplished the tensile test was performed.

RESULTS AND DISCUSSION

Cross-linked silica aerogels are expected to be translucent in the visible range and heavily absorbing in the UV range. The mean UV-Vis transmittance spectra of all aerogel cylinders used in this study are shown in Figure 3 where error bars represent the statistical standard deviation. It can be seen that transmission through each sample increases significantly as the wavelength of the incident beam increases towards the visible region, as expected. The transmittance signal is expected to be affected by the density of the aerogel sample under investigation resulting in a weaker transmitted signal for samples with higher densities. Monolithic aerogel densities of the synthesized samples used in this study fell in the range of 0.317–0.5 gm/cm^3 hence explaining the variations in the transmission spectra.

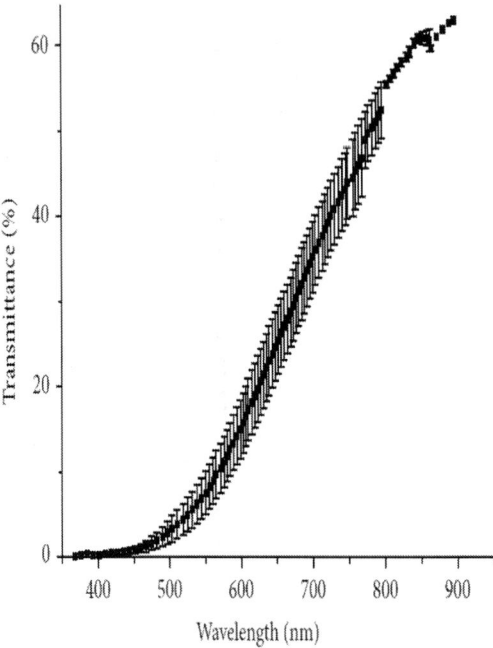

Figure 3: UV-Vis transmission behavior of PCSA monolithic cylinders used in this study prior to the pulverization stage. Monolithic aerogel densities of the synthesized samples fell in the range of 0.317–0.5 gm/cm^3.

The typical room temperature (300 k) and low temperature (77 K) tensile behavior of neat (0% IL) RTV 655 are shown in Figures 4(a) and 4(b), respectively, reflecting a "baseline" behavior and measurement. At room temperature the tensile behavior of 10:1 (polymer: cross-linker) RTV 655 demonstrates a classic elastomeric behavior [40, 41] as expected. The tensile behavior of neat RTV 655 at 77 K demonstrates a significant strengthening and stiffening of the polymer chains [32] with limited amount of elasticity and no noticeable amount of geometrical change in the form of shrinkage. The effect of PCSA microparticle incorporation on the rupture load of impregnated RTV 655 at room temperature is shown in Figure 5 for three separate sample batches measured independently and compared with the behavior of neat RTV 655 measured under similar conditions. For the largest particle sizes (d_1 = 300–425 μm) (Figure 5(a)) a drop of more than a factor of ten in rupture load is noted for the lowest IL (25%) compared to the value for neat RTV 655 leading to a highly nonlinear profile. A less significant change in rupture load (drop by a factor 1.5) is observed when IL is increased from 25% to 75% suggesting that for the larger particle sizes the level of impregnation (loading) plays a small role in the overall mechanical strength of the compound material. In Figure 5(b) the effect of medium-sized particles (d_2 = 180–300 μm) on rupture load shows again a drop in load roughly by a factor of ten between 0% IL and 25% IL and a decrease in rupture load of 0.25 between 25% and 75% with an overall nonlinear behavior again. Finally, the effect of the smallest particle size range (d_3 = 90–180 μm) on rupture load is shown in Figure 5(c) where the decrease in rupture load between 25% and 75% is more pronounced leading to a linear material response overall. In other words, as the particle size shrinks the relationship between rupture load and IL becomes more linear and the overall tensile strength of the material increases, such that as the particle size shrinks by approximately a factor of four, the tensile strength increases by a minimum factor of two and maximum factor of five, never exceeding the rupture load of the neat RTV 655.

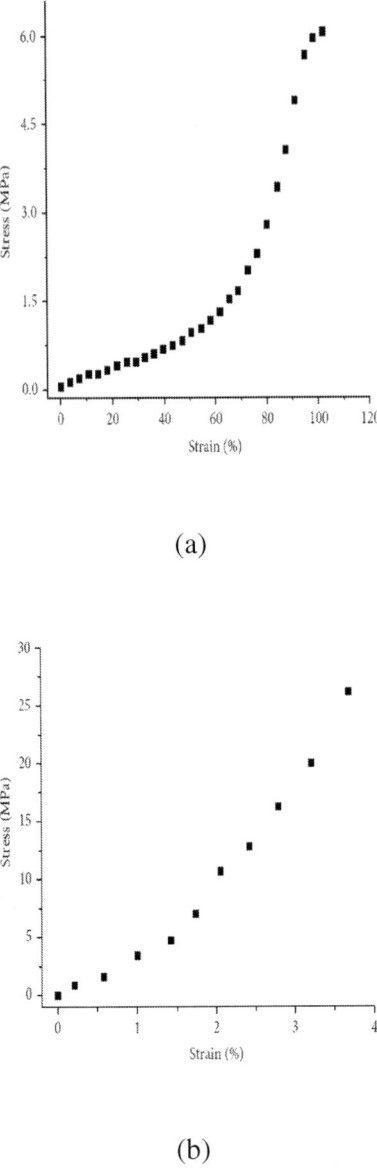

(a)

(b)

Figure 4: Typical tensile behavior of neat (0% IL) RTV 655 at (a) room temperature and (b) 77 K. The stress strain curve of RTV 655 at room temperature follows a classic elastomeric behavior. The increased strength of the polymer at 77 K reflects reduced mobility of the polymer chains and significant stiffening without geometrical changes such as shrinkage.

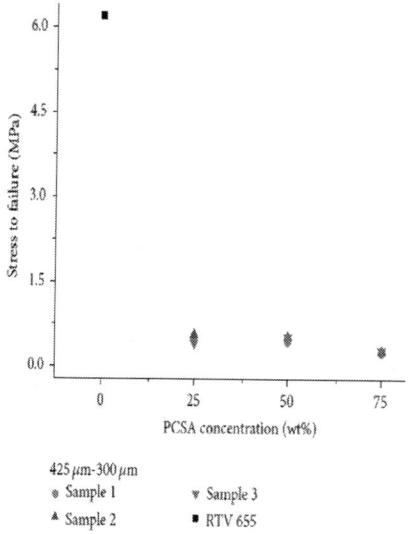

(a)

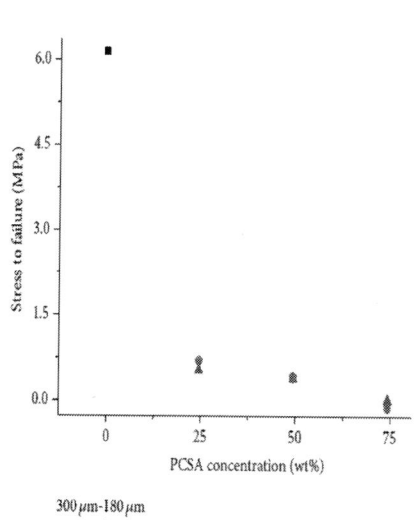

(b)

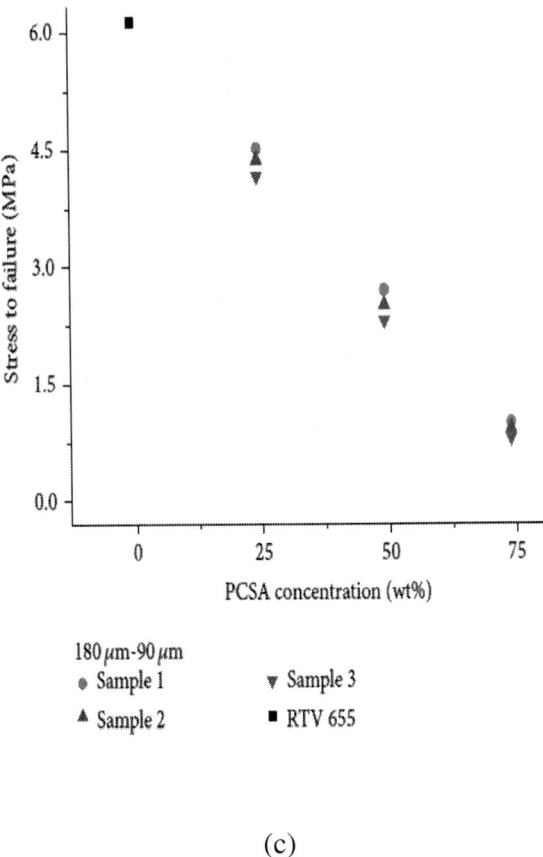

(c)

Figure 5: Comparison of the rupture load of RTV 655+PCSA compound specimens at room temperature with neat (0% IL) RTV 655 samples. Samples were impregnated with three particle size ranges: (a) d_1 = 300–425 µm, (b) d_2 = 180–300 µm, and (c) d_3 = 90–180 µm measured for three separate sample batches. A trend of reduction in the overall mechanical strength of compound material was observed due to incorporation of PCSA particles. However, a least amount of reduction in the yield strength of the compound sample was observed at 25 wt. (%) impregnation with the smallest (d_3 = 90–180 µm) PCSA particles.

The effect of PCSA particle size and particle concentration on rupture load at 77 K is shown in Figure 6 for three separate sample batches. The strong nonlinear behavior observed at room temperature (Figures 5(a) and 5(b)) for particles size ranges d_1 and d_2 at IL 0% to

75% is repeated at 77 K with a linear behavior for the smallest particle size range d_3. Figures 5 and 6 suggest that, by reducing the particle size, at 25% IL the load tolerance of the compound material is comparable with the neat polymer. At IL greater than 25%, however, the load bearing capability of the compound material is significantly reduced and further reduction of particle size has little effect. The rupture load variations seen from one sample batch to another are mainly attributed to particle size variations for each range and exact distribution in the polymeric matrix after the mixing and curing stage has been completed. In both cases (room temperature and 77 K) the smallest particle size range d_3 leads to the highest stress to failure values, consistently for three separate batches. It is hypothesized that this can be improved by further reducing the particle diameters [42].

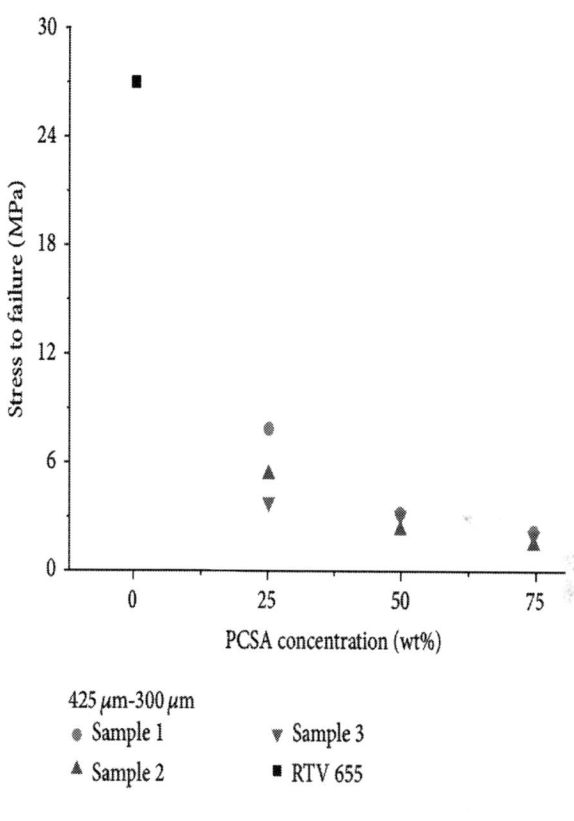

(a)

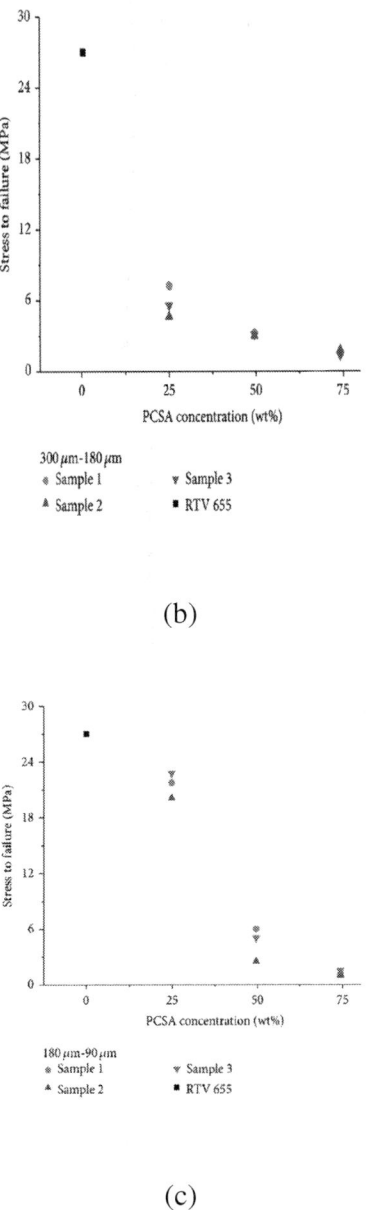

(b)

(c)

Figure 6: Rupture load of RTV 655+PCSA compound specimens at 77 K impregnated with three particle size ranges: (a) d_1 = 300–425 µm, (b) d_2 = 180–300 µm, and (c) d_3 = 90–180 µm measured for three separate sample batches. A similar trend of reduction in the yield strength compared to room temperature measurements was observed. However, a considerable amount of stiffness of both neat RTV 655 and compound specimens was observed.

The effect of the microparticle size and concentration on the overall stress-elongation profile of the compound material was also investigated and compared to the behavior of the neat RTV 655 polymer. A typical room temperature response is shown in Figure 7 where at 25% IL (Figure 7(a)) the compound material shows some degree of elasticity when compared to the neat RTV 655 behavior for all microparticle diameters. As the concentration level increases to 50% (Figure 7(b)) the amount of elasticity drops significantly deviating from the elastomeric behavior and showing stiff nonelastic property with significantly lower stress tolerance. Finally, at 75% IL (Figure 7(c)) the compound material shows stiff and brittle properties with minimum or no elasticity at all. At the 25% IL particle size did play a noticeable and distinguishable role such that some degree of plasticity was observed for each particle size range. As the IL increased, however, the effect of the particle size became insignificant compared to the effect of the IL, gradually transitioning from an elastomeric material to a stronger and tougher material (comparison of slopes). Loading levels in this study are significantly higher than those in a previously reported work [42].

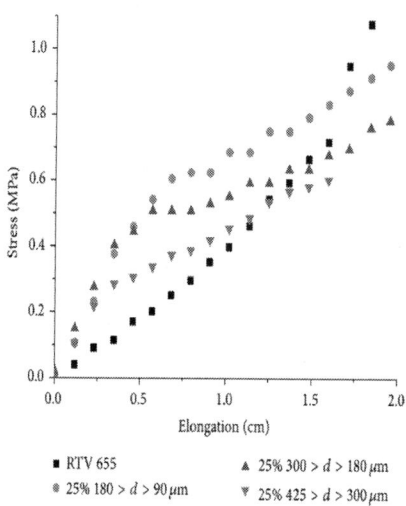

(a)

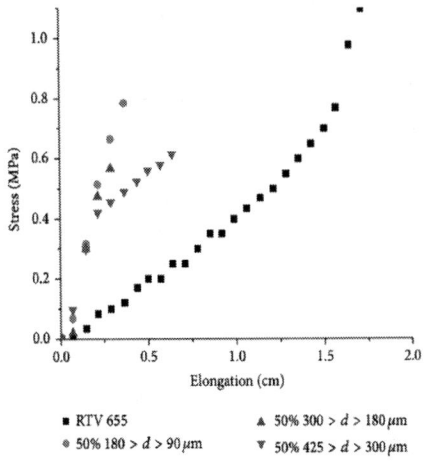

(b)

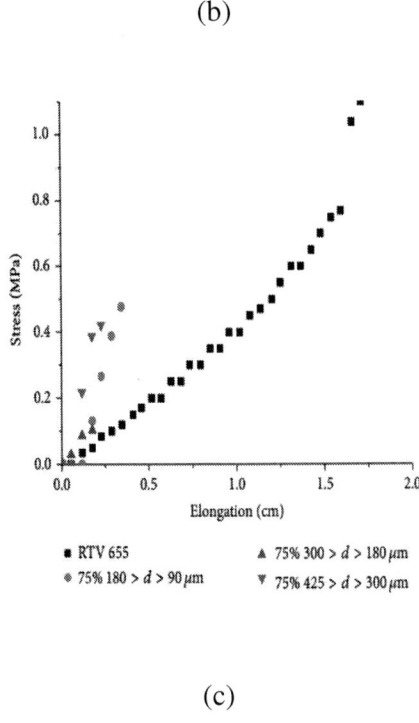

(c)

Figure 7: Effect of PCSA particle size and concentration on the stress-elongation behavior of the compound material at room temperature. At 25% IL (a)

shows a modest deviation from elastomeric behavior in comparison to neat RTV 655. For 50% IL (b) the compound material shows sign of a stiff material while at 75% IL (c) characteristics of a stiff and strong and more brittle material become apparent. The effect of the particle sizes seems less significant compared to the IL.

The stress-elongation response was also evaluated at 77 K for each IL level and all microparticle diameters previously discussed and are shown in Figure 8. For the 25% IL (Figure 8(a)) the compound material has resulted in a less stiff material and overall weaker performance, rupturing at lower load values than the neat RTV 655. As the IL level increases to 50% the trend follows and finally at 75% IL the compound material demonstrates the weakest behavior. Once again, the effect of the particle sizes studied here is less significant than the effect of the doping concentration (IL).

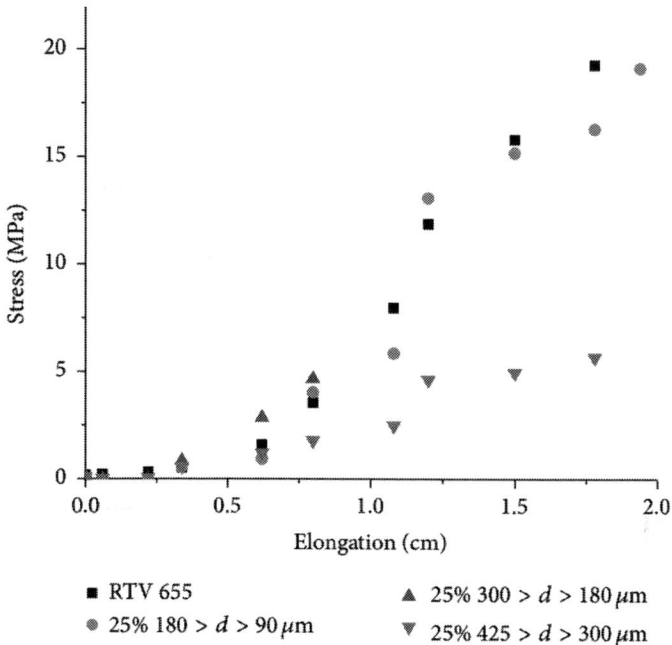

(a)

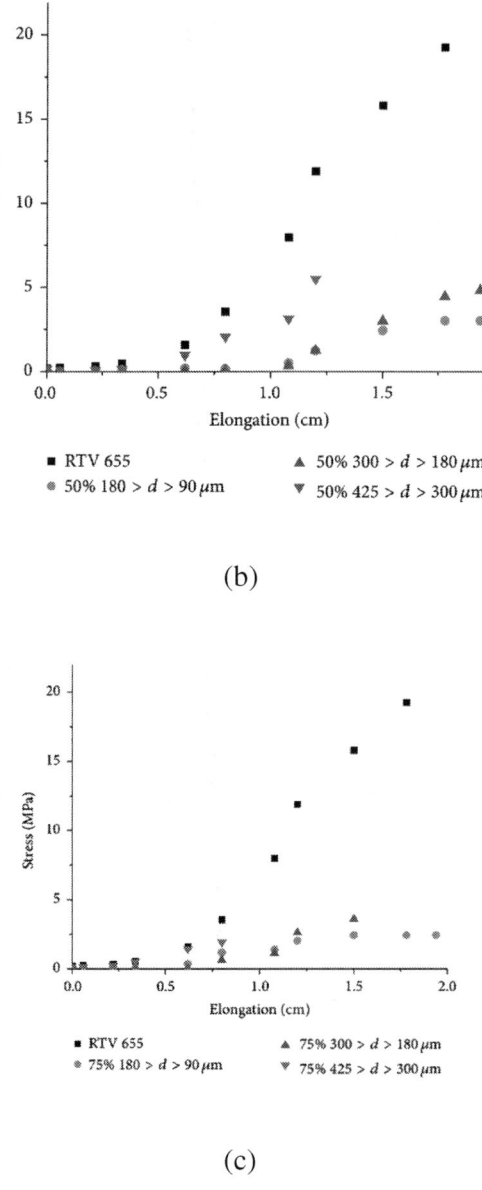

(b)

(c)

Figure 8: Effect of PCSA particle size and concentration on the stress-elongation behavior of the compound material at 77 K. As IL increases from 25% (a) to 50% (b) and finally 75%, (c) the compound material becomes weaker rupturing under lower load values than the neat polymer. Effect of the particle size variation seems less significant than the IL.

The overall reduction of rupture load after introducing PCSA particles into the polymer matrix is not unexpected since the hydrophobic nature of RTV 655 prevents bonding between the PCSA particles and the polymer matrix. It is hypothesized that the PCSA particles are immobilized in the polymer matrix by "physical entrapment." While interface interactions cannot be fully excluded, they are highly unlikely due to (a) the hydrophobic nature of the RTV 655 and (b) observations made during tensile testing. During the tensile testing of the impregnated samples it was observed that, for the larger particles (observed with the aid of magnifiers) as the compound sample was elongated under fixed strain, the elastomeric part of the compound moved at a faster rate and was able to "slide" over the microparticles. It is expected that the same would occur at low temperatures also, but with less travel range available at those temperatures. Direct observations could not be performed at 77 K due to the fact that the samples are positioned in an opaque liquid nitrogen bath that is optically and physically inaccessible. This also prevented the authors from making Poisson ration measurements.

The presence of the microparticles interrupts the formation of long polymer chains and cross-linking that would have otherwise occurred at the microscopic level. Technical challenges associated with sample preparation prevented glass transition temperature measurements for impregnated samples. Scanning electron microscope (SEM) images of the pulverized cross-linked aerogels shown in Figure 9 demonstrate a highly irregular geometry which is the result of the step-pulse pulverizing technique. The geometrical irregularity observed in the microparticles formed affects the magnitude and profile of stress concentrations at the microscopic scale and hence affects the macroscopic mechanical behavior that is measured for each sample. While it is possible to create large quantities of microparticles of varying diameters using this technique, the irregular geometry is thought to contribute significantly towards lowering the rupture load values.

Figure 9: SEM image of pulverized cross-linked silica aerogels showing non-uniform geometries created by using the step-pulse pulverizing technique.

SUMMARY AND CONCLUSIONS

The limitation performance of an aerogel-impregnated RTV 655 elastomer was evaluated for two critical temperatures. Aerogel microparticles were created by means of a step-pulse pulverizing technique that was easy to process and suitable for the formation of large volumes of microparticles. This method however provided little control over the particle geometry and more specifically surface morphology. The majority of microparticles created using this technique had diameters greater than 90 µm, setting the lower limit of the microparticle size distributions used in this study. It was discovered that the maximum impregnation and loading level that resulted in complete curing and encapsulation was discovered to be 75 wt%.

The effect of size and concentration of polyurea cross-linked silica aerogel microparticles on the tensile behavior of the elastomer RTV 655 was fully characterized at two critical temperatures of room temperature and 77 K. As the concentration of PCSA microparticles increased from 0 to 75 wt% the tensile behavior of the compound material transitioned from an elastic behavior to a stiff and tough material. For IL of 25 and 50 wt% the effect of microparticle diameter on overall material response was identifiable. At IL 75 wt% however the concentration level dominated the material behavior with little influence from the particle diameter variations. Overall, the size of the PCSA particles impacted the rupture load values the most while the concentration of the PCSA particles in the RTV 655 matrix dominated the stress-elongation behavior. It is expected that by further reducing the particle sizes significant increase in the rupture load values will be seen.

ACKNOWLEDGMENTS

The authors would like to acknowledge the financial support from NASA TN EPSCOR NNX10AQ71A. The authors would also like to thank John Daffron from the Department of Physics Instrument Shop at the University of Memphis as well as Robert Hewitt at the University of Memphis for technical support.

REFERENCES

1.   F. Sabri, J. A. Cole, M. C. Scarbrough, and N. Leventis, "Investigation of polyurea-crosslinked silica aerogels as a neuronal scaffold: a pilot study," PLoS ONE, vol. 7, no. 3, Article ID e33242, 2012.

2.   T. Mehling, I. Smirnova, U. Guenther, and R. H. H. Neubert, "Polysaccharide-based aerogels as drug carriers," Journal of Non-Crystalline Solids, vol. 355, no. 50-51, pp. 2472–2479, 2009.

3.   Y. K. Li, M. J. Chou, T.-Y. Wu, T.-R. Jinn, and Y. W. Chen-Yang, "A novel method for preparing a protein-encapsulated bioaerogel: using a red fluorescent protein as a model," Acta Biomaterialia, vol. 4, no. 3, pp. 725–732, 2008.

4.   M. D. Frogley, D. Ravich, and H. D. Wagner, "Mechanical properties of carbon nanoparticle-reinforced elastomers," Composites Science and Technology, vol. 63, no. 11, pp. 1647–1654, 2003.

5.   S.-Y. Fu and B. Lauke, "Fracture resistance of unfilled and calcite-particle-filled ABS composites reinforced by short glass fibers (SGF) under impact load," Composites Part A: Applied Science and Manufacturing, vol. 29, no. 5-6, pp. 631–641, 1998.

6.   A. C. Moloney, H. H. Kausch, T. Kaiser, and H. R. Beer, "Parameters determining the strength and toughness of particulate filled epoxide resins," Journal of Materials Science, vol. 22, no. 2, pp. 381–393, 1987.

7.   A. J. Kinloch and R. J. Young, "Fracture behavior of polymers," British Polymer Journal, vol. 16, no. 2, p. 114, 1984.

8.   M. E. J. Dekkers and D. Heikens, "The effect of interfacial adhesion on the tensile behavior of polystyrene–glass-bead composites," Journal of Applied Polymer Science, vol. 28, no. 12, pp. 3809–3815, 1983.

9.   S.-Y. Fu and B. Lauke, "Characterization of tensile behaviour of hybrid short glass fibre/calcite particle/ABS composites," Composites A: Applied Science and Manufacturing, vol. 29, no. 5-6, pp. 575–583, 1998.

10.   K. C. Radford, "The mechanical properties of an epoxy resin with a second phase dispersion," Journal of Materials Science, vol. 6, no. 10, pp. 1286–1291, 1971.

11.   Y. Nakamura, M. Yamaguchi, M. Okubo, and T. Matsumoto, "Effects of particle size on mechanical and impact properties of epoxy resin filled with spherical silica," Journal of Applied Polymer Science, vol. 45, no. 7, pp. 1281–1289, 1992.

12.   L. Nicolais and L. Nicodemo, "Effect of particles shape on tensile properties of glassy thermoplastic composites," International Journal of Polymeric Materials, vol. 3, no. 3, pp. 229–243, 1974.

13.   T. K. Jayasree and P. Predeep, "Effect of fillers on mechanical properties of dynamically crosslinked styrene butadiene rubber/ high density polyethylene blends," Journal of Elastomers and Plastics, vol. 40, no. 2, pp. 127–146, 2008.

14. K. W. Bills Jr., K. H. Sweeny, and F. S. Salcedo, "The tensile properties of highly filled polymers. Effect of filler concentrations," Journal of Applied Polymer Science, vol. 4, no. 12, pp. 259–268, 1960.

15. C.-H. Hsueh, "Effects of aspect ratios of ellipsoidal inclusions on elastic stress transfer of ceramic composites," Journal of the American Ceramic Society, vol. 72, no. 2, pp. 344–347, 1989.

16. R. J. Young and P. W. R. Beaumont, "Failure of brittle polymers by slow crack growth—Part 3 Effect of composition upon the fracture of silica particle-filled epoxy resin composites," Journal of Materials Science, vol. 12, no. 4, pp. 684–692, 1977.

17. B. Pukanszky and G. Voros, "Mechanism of interfacial interactions in particulate filled composites,"Composite Interfaces, vol. 1, no. 5, pp. 411–427, 1993.

18. Y. Ou, F. Yang, and Z.-Z. Yu, "A new conception on the toughness of nylon 6/silica nanocomposite prepared via in situ polymerization," Journal of Polymer Science, Part B: Polymer Physics, vol. 36, no. 5, pp. 789–795, 1998.

19. Z. K. Zhu, Y. Yang, J. Yin, and Z. N. Qim, "Preparation and properties of organosoluble polyimide/silica hybrid materials by sol–gel process," Journal of Applied Polymer Science, vol. 73, no. 14, pp. 2977–2984, 1999.

20. E. Reynaud, T. Jouen, C. Gauthier, G. Vigier, and J. Varlet, "Nanofillers in polymeric matrix: a study on silica reinforced PA6," Polymer, vol. 42, no. 21, pp. 8759–8768, 2001.

21. B. E. Yoldas, M. J. Annen, and J. Bostaph, "Chemical engineering of aerogel morphology formed under nonsupercritical conditions for thermal insulation," Chemistry of Materials, vol. 12, no. 8, pp. 2475–2484, 2000.

22. B. Wolff, G. Seybold, and F. E. Krueckau, "Thermal insulators having density 0.1 to 0.4 g/cm³, and their manufacture BASF-G," European Patent Applications 0340707, 2010.

23. F. Sabri, J. Marchetta, and K. M. Smith, "Thermal conductivity studies of a polyurea cross-linked silica aerogel-RTV 655 compound for cryogenic propellant tank applications in space," Acta Astronautica, vol. 91, pp. 173–179, 2013.

24. B. E. Scholtens, J. E. Fesmire, J. P. Sass, S. D. Augustynowicz, and K. W. Heckle, "Cryogenic thermal performance testing of bulk-fill and aerogel insulation materials," in Proceedings of the AIP Conference, pp. 152–159, Chattanooga, Tenn, USA, 2008.

25. M. K. Williams, T. M. Smith, and J. E. Fesmire, "Aerogel/polymer composite materials," US patent no. US7790787, 2010.

26. K. S. Novak, C. J. Phillips, G. C. Burir, E. T. Sunada, and M. T. Pauken, "Development of a thermal control architecture for the Mars Exploration Rovers," in Proceedings of the AIP Conference, pp. 194–205, 2003.

27. A. Katti, N. Shimpi, S. Roy et al., "Chemical, physical, and mechanical characterization of isocyanate cross-linked amine-modified silica aerogels," Chemistry of Materials, vol. 18, no. 2, pp. 285–296, 2006.

28. H. Luo, G. Churu, E. F. Fabrizio et al., "Synthesis and characterization of the physical, chemical and mechanical properties of isocyanate-crosslinked vanadia aerogels," Journal of Sol-Gel Science and Technology, vol. 48, no. 1-2, pp. 113–134, 2008.

29. Y. Jia, S. Sun, L. Liu, S. Xue, and G. Zhao, "Investigation of computer-aided engineering of silicone rubber vulcanizing (I)—vulcanization degree calculation based on temperature field analysis,"Polymer, vol. 44, no. 1, pp. 319–326, 2002.

30. Q. G. Gu and Q. L. Zhou, "Preparation of high strength and optically transparent silicone rubber,"European Polymer Journal, vol. 34, no. 11, pp. 1727–1733, 1998.

31. M. Patel, A. R. Skinner, and R. S. Maxwell, "Sensitivity of condensation cured polysiloxane rubbers to sealed and open-to air thermal ageing regimes," Polymer Testing, vol. 24, no. 5, pp. 663–668, 2005.

32. D. A. Wigley, Mechanical Properties of Materials at Low Temperatures, Plenum Press, New York, NY, USA, 1971.

33. M. A. B. Meador, S. L. Vivod, L. McCorkle et al., "Reinforcing polymer cross-linked aerogels with carbon nanofibers," Journal of Materials Chemistry, vol. 18, no. 16, pp. 1843–1852, 2008.

34. S. Mulik, C. Sotiriou-Leventis, and N. Leventis, "Macroporous electrically conducting carbon networks by pyrolysis of

isocyanate-cross-linked resorcinol-formaldehyde aerogels," Chemistry of Materials, vol. 20, no. 22, pp. 6985–6997, 2008.

35. M. A. B. Meador, L. A. Capadona, L. McCorkle, D. S. Papadopoulos, and N. Leventis, "Structure-property relationships in porous 3D nanostructures as a function of preparation conditions: isocyanate cross-linked silica aerogels," Chemistry of Materials, vol. 19, no. 9, pp. 2247–2260, 2007.

36. F. Sabri, T. Werhner, J. Hoskins et al., "Thin film surface treatments for lowering dust adhesion on Mars Rover calibration targets," Advances in Space Research, vol. 41, no. 1, pp. 118–128, 2008.

37. J. K. Lee, G. L. Gould, and W. Rhine, "Polyurea based aerogel for a high performance thermal insulation material," Journal of Sol-Gel Science and Technology, vol. 49, no. 2, pp. 209–220, 2009.

38. F. Sabri, N. Leventis, J. Hoskins et al., "Spectroscopic evaluation of polyurea crosslinked aerogels, as a substitute for RTV-based chromatic calibration targets for spacecraft," Advances in Space Research, vol. 47, no. 3, pp. 419–427, 2011.

39. J. L. Goudie and T. P. Collins, "Development and evaluation of an improved RTV coating for outdoor insulation," in Proceedings of the IEEE International Symposium on Electrical Insulation, pp. 475–479, Indianapolis, Ind, USA, September 2004.

40. W. D. Callister Jr. and D. G. Rethwisch, Material Science and Engineering: An Introduction, John Wiley & Sons, Hoboken, NJ, USA, 8th edition, 2010.

41. F. Schneider, T. Fellner, J. Wilde, and U. Wallrabe, "Mechanical properties of silicones for MEMS,"Journal of Micromechanics and Microengineering, vol. 18, no. 6, Article ID 065008, 2008.

42. Z. Wang, J. Liu, S. Wu, W. Wang, and L. Zhang, "Novel percolation phenomena and mechanism of strengthening elastomers by nanofillers," Physical Chemistry, Chemical Physics, vol. 12, no. 12, pp. 3014–3030, 2010.

Chapter 4

Mechanical Behaviors of Friction Stir Spot Welded Joints of Dissimilar Ferrous Alloys under Opening-Dominant Combined Loads

Md. Abu Mowazzem Hossain[1], Md. Tariqul Hasan[1], Sung-Tae Hong[1], Michael Miles[2], Hoon-Hwe Cho[3,] and Heung Nam Han[3]

[1]School of Mechanical Engineering, University of Ulsan, Ulsan 680-749, Republic of Korea

[2]Department of Manufacturing Engineering Technology, Brigham Young University, Provo, UT 84602-9601, USA

[3]Department of Materials Science & Engineering and Center for Iron & Steel Research, RIAM, Seoul National University, Seoul 151-744, Republic of Korea

ABSTRACT

Mechanical properties and failure behaviors of friction stir spot welded (FSSW) joints of two dissimilar ferrous alloys, cold-rolled carbon steel (SPCC) and 409L stainless steel (SUS 409L), are investigated under opening-dominant combined loads. The texture of dissimilar FSSW joints depends on the upper sheet material. The failure contours for the FSSW joints under combined loads are constructed in terms of the axial load and shear load by modifying existing failure criteria for resistance spot welds. The shape of the failure contour also depends on the upper sheet material. The failure contours are nearly elliptic in shape when the upper sheet is SPCC and are relatively straight lines when the upper sheet is SUS 409L.

INTRODUCTION

Friction stir spot welding (FSSW) is a solid-state joining process that originated from friction stir welding (FSW), a technique patented in 1991 by TWI [1]. Compared with conventional resistance spot welding (RSW), FSSW provides several technical advantages including low energy requirements, fewer problems related to cracking and porosity, less residual stress, and a smaller heat affected zone (HAZ) [2, 3]. Also, an additional interesting technical advantage over conventional RSW is that FSSW can easily join dissimilar metal alloys.

Even though RSW has been the technique most commonly used in the automotive industries for joining ferrous alloys, RSW is also known to be problematic for several specific ferrous alloys [4]. Moreover, the use of dissimilar ferrous alloys is increasing in the automotive industry to improve the crashworthiness of the automobile structure without significantly increasing weight and manufacturing cost. It is clear that RSW of dissimilar ferrous alloys can be extremely difficult due to the different physical, chemical, and mechanical properties of the base metals [5].

Due to FSSW's technical advantages, including its ability to join dissimilar alloys easily, interest has increased in FSSW as a substitute for RSW in joining similar or dissimilar ferrous alloys in automotive applications. However, even though many pioneering works on FSSW of ferrous alloys have been conducted over the last decade, substituting FSSW for the RSW commonly used in automotive applications will require further investigation. Feng et al. [6] performed an introductory study that examined the feasibility of FSSW of AHSS steel; this study suggested that the mechanical strength of the FSSW joint improves as the width of the bonding ligament increases. Baek et al. [7] examined the effect of tool penetration depth on the microstructures and mechanical properties of FSSW joints of low-carbon steel. Their results showed that the tensile shear strength of low-carbon steel FSSW joint increases as the tool penetration depth increases. Hovanski et al. [8] investigated the microstructure and mechanical properties of FSSW joints of hot-stamped boron steel, as well as their failure mechanism: cracking initiated at the interface of the upper and lower sheets and then propagated along the thin ferritic region within weld nugget. Miles et al. [9] investigated the effect of tool wear on joint strength using tools composed of polycrystalline cubic boron nitride (PCBN) and tungsten rhenium (W-Re), where tools with a greater proportion of PCBN were found to provide the best combination of joint strength and wear resistance. Sun et al. [10] experimentally investigated the failure behavior of FSSW joints of mild steel, suggesting based on the result of shear tensile tests, that FSSW joints undergoing fracture by the plug failure mode have higher shear tensile strength compared to FSSW joints undergoing fracture by the interfacial failure mode.

A critical mechanical property of a joint in an automobile body structure is its failure behavior. For RSW, research has been conducted on joint failure, including the development of different fracture models formulated in terms of the local loads acting on the spot welds, as well as in terms of the appropriate strengths of the weld. Chao [11] developed a failure criterion for spot welds and performed strength tests using cross tension and lap shear samples made of high-strength steel. Wung [12] and Wung et al. [13] investigated the failure of spot welds under in-plane torsion and proposed a force-based failure criterion. Radaj [14], Radaj and Zhang [15], and Zhang [16] adopted a fracture mechanics approach and provided a very detailed description of the stress distribution around a weld nugget. Radaj [14] showed that the fatigue strength of spot-welded joints can be assessed on the basis of the local stress state at the weld spot edge. Radaj and Zhang [15] developed the relations between notch stress and crack stress intensity in the case of plane shear loading (mode II) for the elliptical hole and the blunt crack. Zhang [16] derived approximate stress formulas of structural stress and notch stress for a newly proposed multiaxial spot weld specimen that enables a spot weld to be tested under combined loads ranging from pure shear to pure tension.

The failure behavior of RSW joints under combined or multiaxial loads also has been investigated by many researchers. The failure behavior of a joint under combined loads is typically important in the structural durability and crash safety of the automobile structure. Lee et al. [17] performed a failure test of RSW joints in U-shaped specimens under combined shear and tension loads and proposed an ultimate strength model to fit their experimental results. Lin et al. [18, 19] analyzed the failure mechanism of spot welds in square-cup specimens made from mild steel and HSLA steel under combined loads. They proposed a quadratic-form engineering failure criterion in terms of the normalized axial and shear loads with consideration of the sheet thickness and the nugget radius under combined loads. More recently, a different failure criterion in terms of the axial and shear loads has been suggested by Song and Huh [20] to predict the failure behavior of RSW joints.

However, the failure behavior of FSSW joints under combined or multiaxial loads has been rarely investigated until recently, even though the failure behavior of FSSW joints can differ from that of RSW joints [21] due to the different joining mechanisms of FSSW and RSW. In the present study, the failure behavior of FSSW joints of two dissimilar ferrous alloys under opening-dominant combined loads was investigated experimentally. Based on the experimental result, failure contours for the FSSW joints were constructed in terms of the axial and shear loads. A failure criterion, which was originally developed for RSW joints, was then modified to describe the experimental failure contours of the FSSW joints.

EXPERIMENTAL

The two dissimilar ferrous alloys used in the present study were 1.2 mm thick steel sheets of cold-rolled carbon steel (SPCC) and 409L stainless steel (SUS); Table 1 lists the nominal chemical compositions and mechanical properties of these alloys. Cross-sectional samples for microstructure analysis were first prepared from FSSW joints of four different material combinations. The welding was carried out using a convex scrolled shoulder tool made of polycrystalline cubic boron nitride (PCBN)-based composite. The process parameters and tool geometry used to fabricate the FSSW joints are listed in Table 2. During the FSSW process, an argon shroud was introduced using a gas cup located around the tool to minimize the surface oxidation of the joint. The samples were mechanically ground and electrolytically polished in a solution of 10 mL perchloric acid + 90 mL ethanol using a Struers Lectropol-5 electrolytic polisher. High-resolution EBSD studies were performed using a Jeol JSM6500F FE-SEM equipped with an HKL Channel 5 EBSD system. The accelerating voltage was 20 kV, the probe current was 4 nA, and the working distance was 15 mm, with the sample stage tilted by 70°. The camera resolution was 1000×800 pixels in the operation of 8×8 binning. The mapping grid was a regular square with 0.7 μm steps. The limits of the low-angle boundaries (LABs) and high-angle boundaries (HABs) were, respectively, set to 2° and 15°. The grain size was measured using the linear intercept method. The hardness profiles in the stir zone and surrounding regions of the FSSW joints were also measured as a function of distance from the weld center.

Table 1: The chemical compositions provided by the manufacturer and mechanical properties of SPCC and SUS 409L

Chemical compositions (wt%)

	C	Mn	P	Si	S	S-AL	Fe
SPCC	0.0361	0.205	0.015	0.019	0.006	0.037	At balance

(a)

Chemical compositions (wt%)

	C	Cr	Mn	P	Si	S	Ni	Ti	Fe
SUS 409L	≤0.03	11.44	≤1.0	≤0.04	≤1.0	≤0.03	≤0.08	≤0.75	At balance

(b)

(c)

Mechanical properties

	Tensile strength (MPa)	Yield strength (MPa)	Elongation at fracture (%)
SPCC	316.8	163.8	46
SUS 409L	494	236	36

Table 2: Friction stir spot welding parameters and tool geometry used in the experiments

Process parameters

Rotationb (rpm)	Plunging rate (mm/min)	Depth (mm)	Control mode	Dwell time (sec)
1400	8	1.45	Servo control	2

Tool geometry

Shoulder diameter (mm)	Pin diameter (mm)	Pin length (mm)
36.8	5.7	1

Square-cup specimens made of SPCC and 409L SUS were then fabricated; these specimens were then friction stir spot welded as depicted schematically in Figure 1, with four different material combinations: SPCC/SPCC, SPCC (top)/SUS (bottom), SUS (top)/SPCC (bottom), and SUS/SUS. Once again, the welding was carried out using a convex scrolled shoulder tool made of PCBN-based composite with the process parameters listed in Table 2. Note that the four corners of each square-cup specimen were arc-welded to ensure adequate stiffness of the cup specimen under opening-dominant combined loads and to guarantee a relatively uniform loading along the circumference of spot weld.

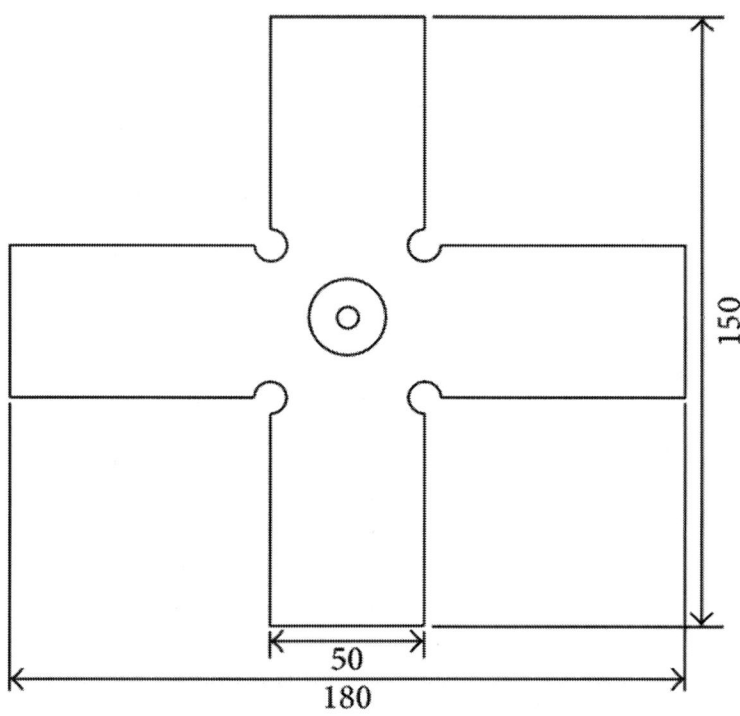

(a)

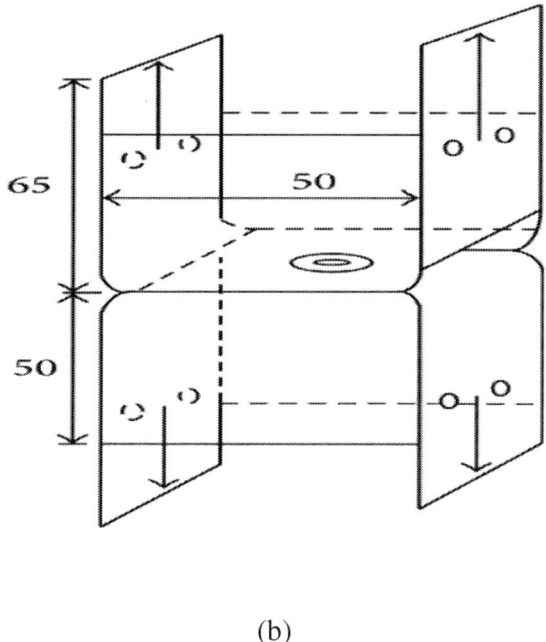

(b)

Figure 1: Schematics of fabricated and welded specimens: (a) top view of un-folded square-cup specimen and (b) folded square-cup specimen with FSSW (dimensions in mm).

In order to impose a combined load on the FSSW joint, four fixture sets were designed with different inclined loading angles ϕ of 0o, 15o, 22o, and 30o (Figure 2), where ϕ is the angle between the load application line and the center line of the FSSW joint. In each experimental setup, the welded squarecup specimens were mounted to the fixture set by means of bolts through the specimen holes. The reinforcement plates were attached to the specimen to prevent plastic deformation near the specimen holes during loading. Using the inclined loading angle ϕ, the applied load$F$ can be simply decomposed into the axial load F_N and the shear load F_S as

$$F_N = F \cos \phi,$$
$$F_S = F \sin \phi.$$

(1)

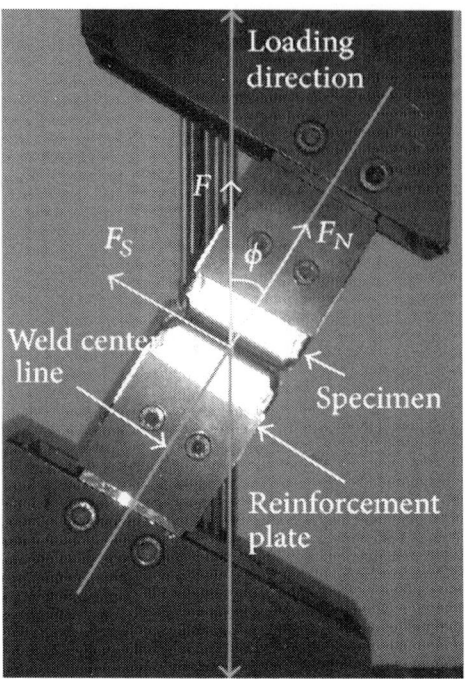

Figure 2: Experimental setup for $\phi=30°$.

Quasi-static tests of the FSSW joints under combined loads were conducted using a universal testing machine with a displacement rate of 2 mm/min along the load application line. During each test the load and displacement were recorded as functions of time. In general, the test was terminated when the FSSW joint was completely separated. The cross section through the weld center of each failed FSSW joint was examined using an optical microscope to understand the failure behavior for each loading condition.

RESULTS AND DISCUSSION

Microstructures

FSSW joints without visible macroscopic defects were successfully produced with the selected process parameters for all four material combinations studied, as evidenced by a representative cross-sectional macrograph of SUS/SPCC FSSW joint in Figure 3 [21]. The indentation profile of the joint reflects the shape of the pin and the convex shoulder of the tool. The surface of the tool was examined visually after each weld. There was no significant tool wear observed throughout the entire set of experiments conducted for the present study.

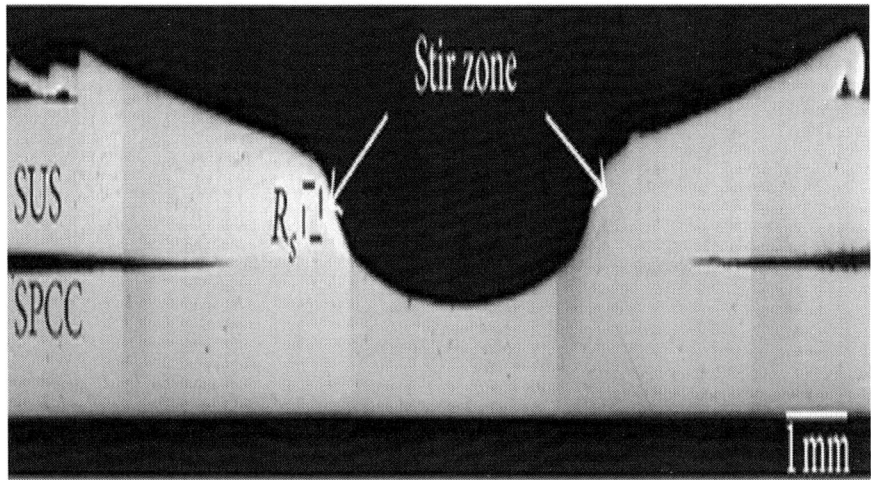

Figure 3: A cross-sectional macrograph of a SUS/SPCC FSSW joint.

The mixing of materials in the stir zone of dissimilar FSSW joints (SPCC/SUS and SUS/SPCC) was already presented in our previous study [21]. We present the mixing of materials in the stir zone here once again just for the completeness of the study. The mixing of materials in the stir zone of dissimilar FSSW joints depends on the material combination (Figures 4(a) and 4(b)) [21]. For the SPCC/SUS FSSW joint, the lower SUS sheet material was pulled upward and mixed into the upper SPCC sheet material as layers (Figure 4(c)). On the other hand, for the SUS/ SPCC FSSW joint, a small portion of SPCC in the lower sheet was pulled upward into the SUS in the upper sheet and relatively little mechanical mixing occurred between the SPCC and the SUS (Figure4 (d)). The result of a line scan of the Cr content along the interface of the region R1 confirms the mechanical mixing between the SPCC and SUS as shown in Figure 4(e).

(a)

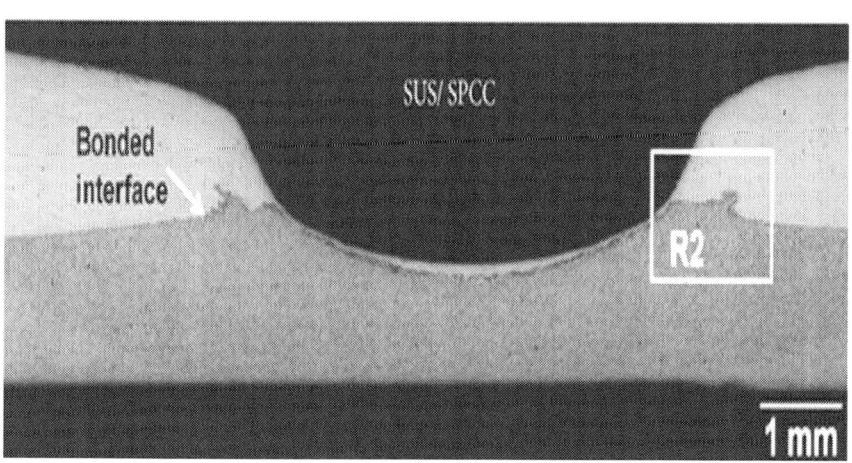

(b)

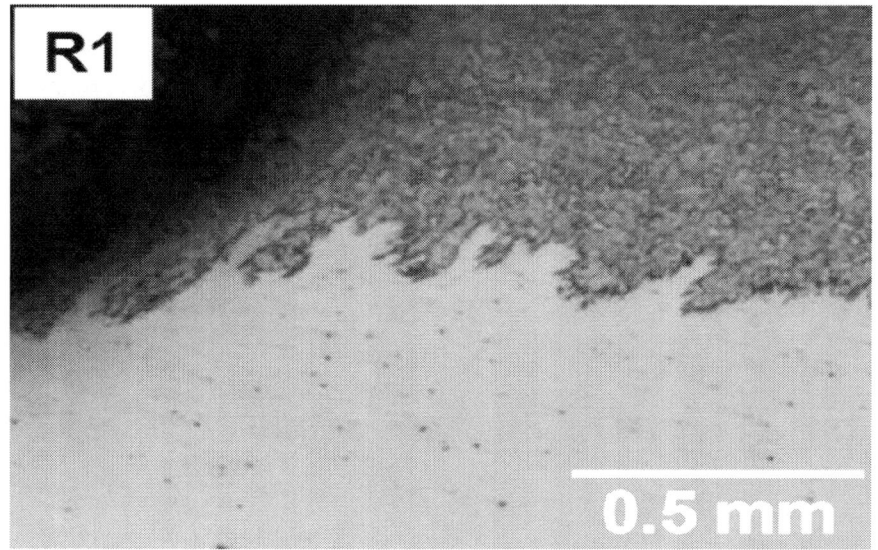

(c)

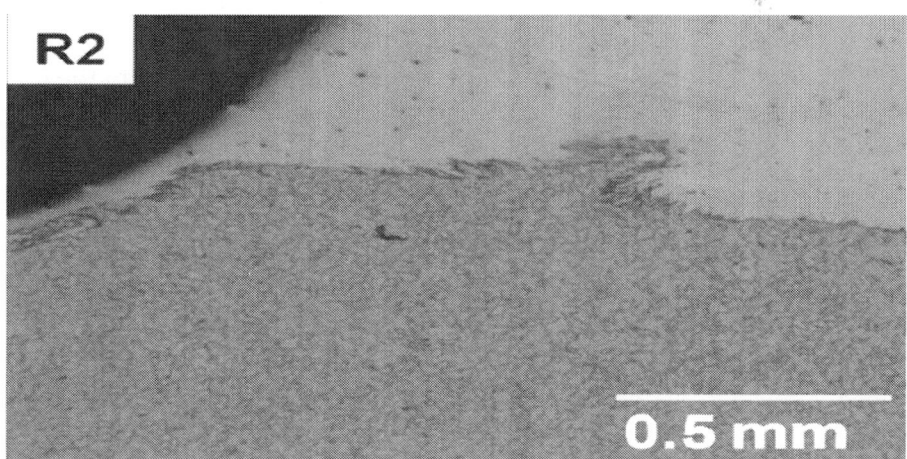

(d)

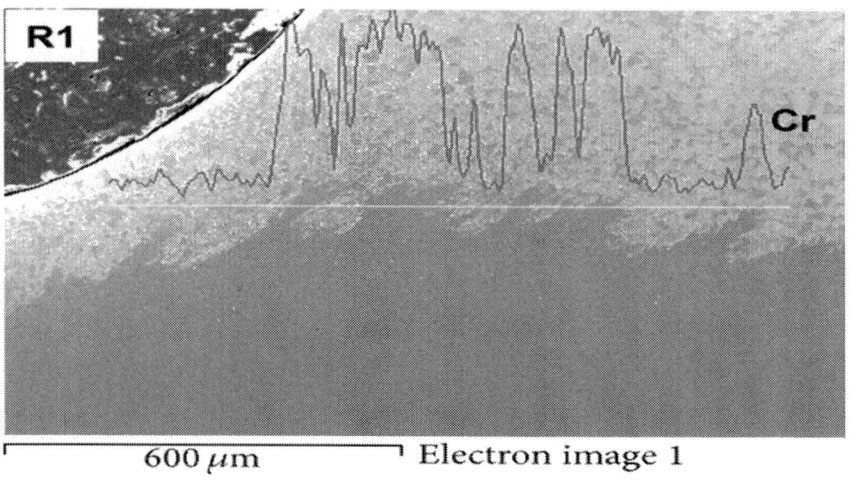

(e)

Figure 4: Optical micrographs of (a) SPCC/SUS and (b) SUS/SPCC FSSW joints; (c) region R1 in (a) and (d) region R2 in (b); (e) Cr distribution profile in region R1 [21].

The orientation maps of the stir zone underneath the root of the pin (R_s in Figure 3 and similar locations for FSSW joints of other material combinations) show that considerably finer homogeneous grains developed in the stir zone in comparison with the base materials as shown in Figures 5(a)–5(f). The average grain size of the base material was approximately 13.2 μm for SPCC and 27.4 μm for SUS. Note that the grain sizes of the stir zone differed somewhat among the four different material combinations. The average grain size of the stir zone of the SPCC/SPCC joint, 10.74 μm, was somewhat larger than that of the SUS/SUS joint, 6.89 μm. This may be explained by the different phase transformation temperatures of the base materials. A further discussion on the phase transformation of the base materials during FSSW is beyond the scope of the present study and will be discussed elsewhere.

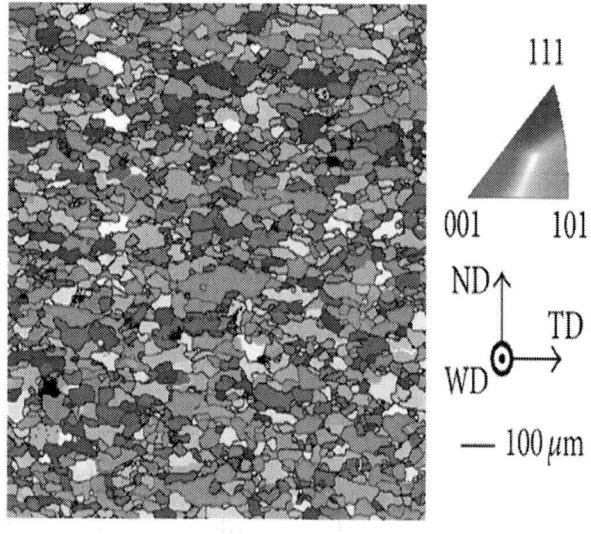

(a)

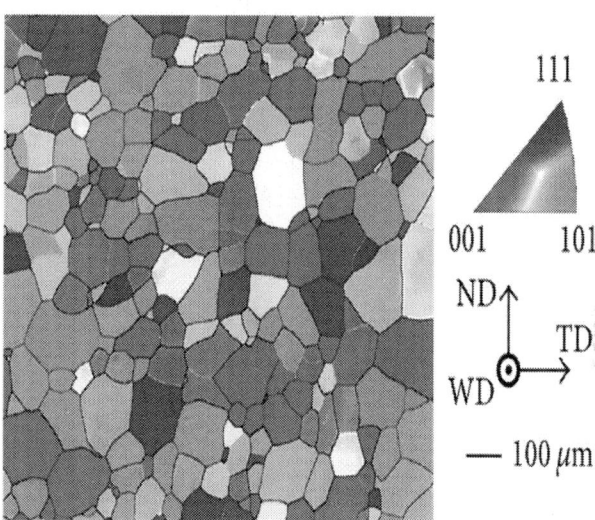

(b)

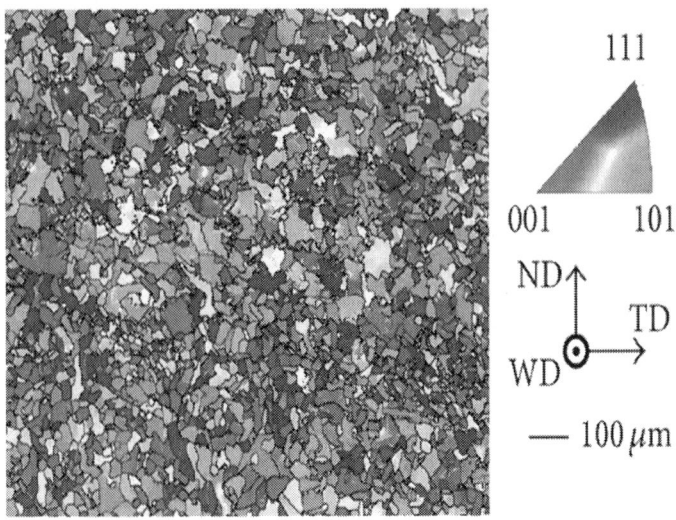

(c)

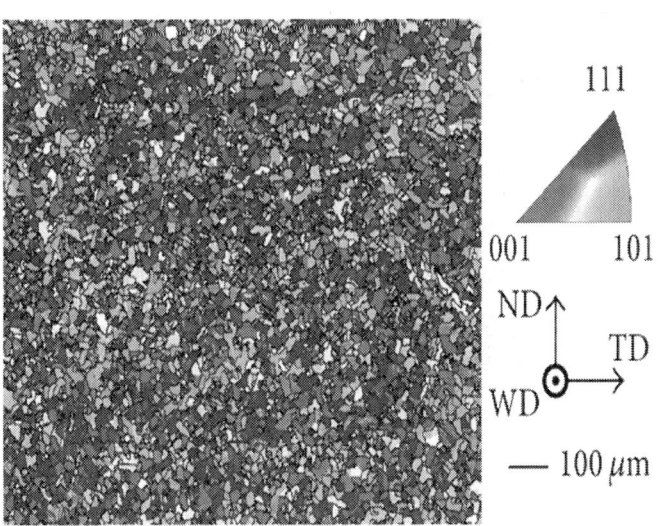

(d)

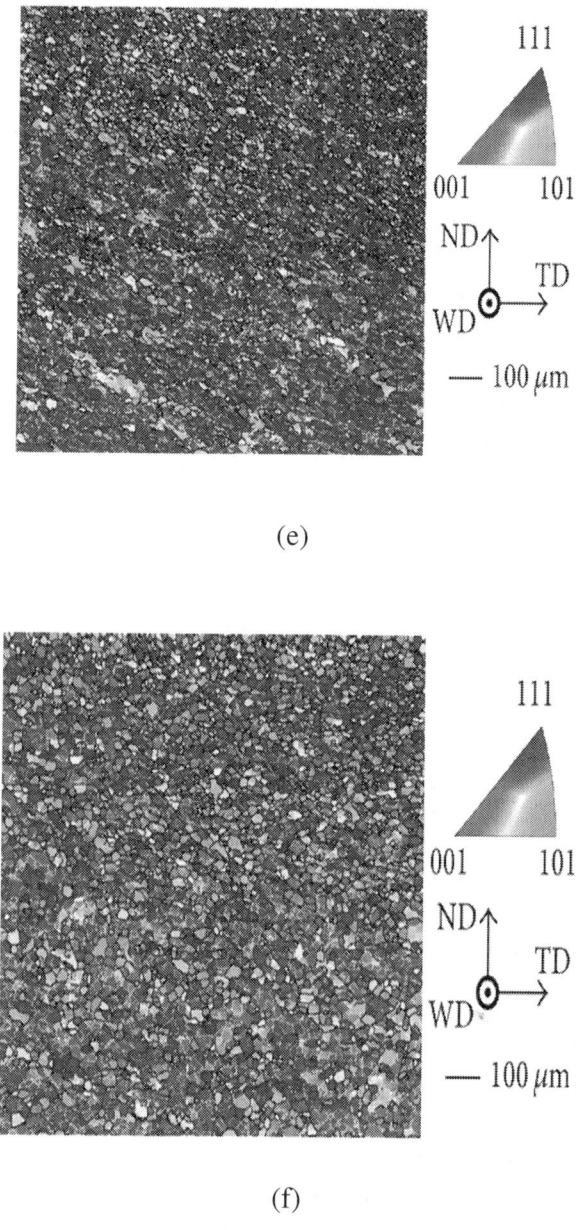

(e)

(f)

Figure 5: Orientation maps of the base material in (a) SPCC and (b) SUS and the stir zone in (c) SPCC/SPCC, (d) SPCC/SUS, (e) SUS/SPCC, and (f) SUS/ SUS, respectively; WD, TD, and ND, respectively, correspond to the welding, transversal, and normal directions.

It is interesting to note that the texture of dissimilar FSSW joints strongly depends on the upper sheet material. Even though the 8.76 µm average grain size of the SPCC/SUS joint is only slightly larger than that of the SUS/SPCC joint, 8.39 µm, as shown in Figures 5(e) and 5(f), the texture of the stir zone of the SPCC/SUS joint is almost random (a very weak shear texture) and differs considerably from that of the SUS/SPCC joint as shown in Figures 6(b) and 6(d). In the stir zone of the SUS/SPCC joint, a clear shear texture developed, similar to the texture of the SUS/SUS joint as shown in Figures 6(c) and 6(d). The dependence of the microstructural characteristics on the upper sheet material is probably due to that only the upper sheet material contacts the tool directly during most of the FSSW process.

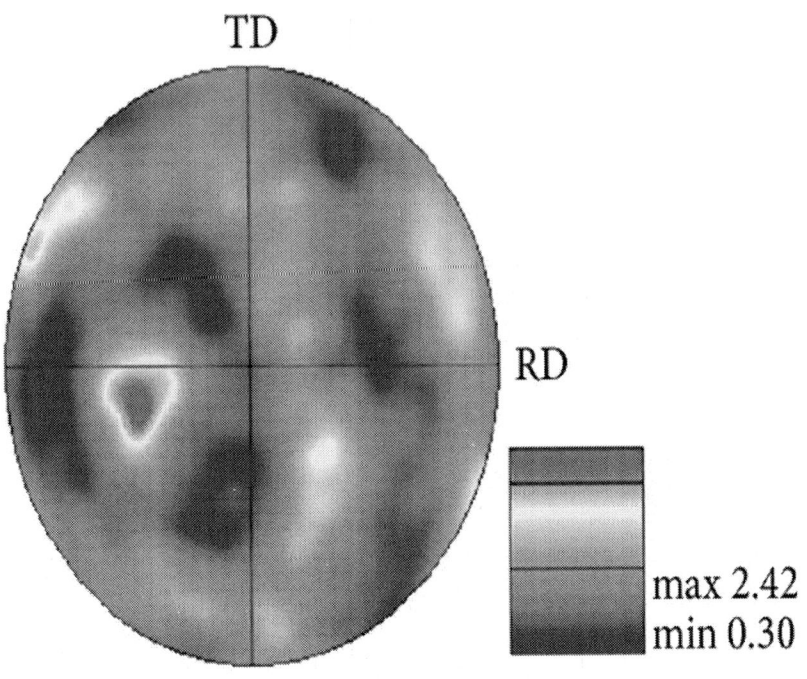

max 2.42
min 0.30

(a)

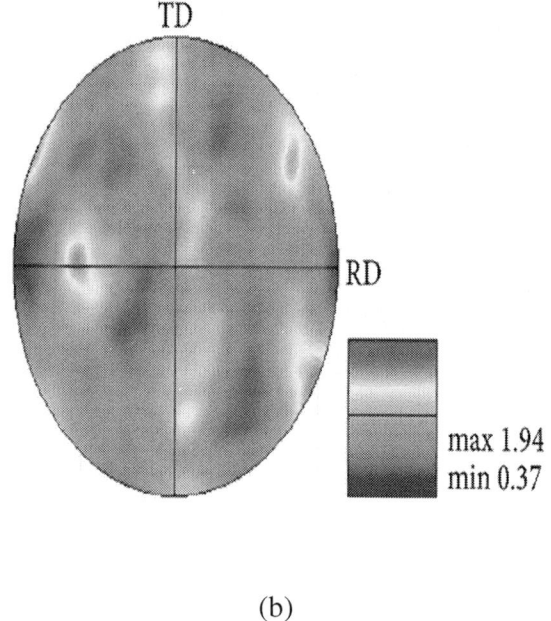

max 1.94
min 0.37

(b)

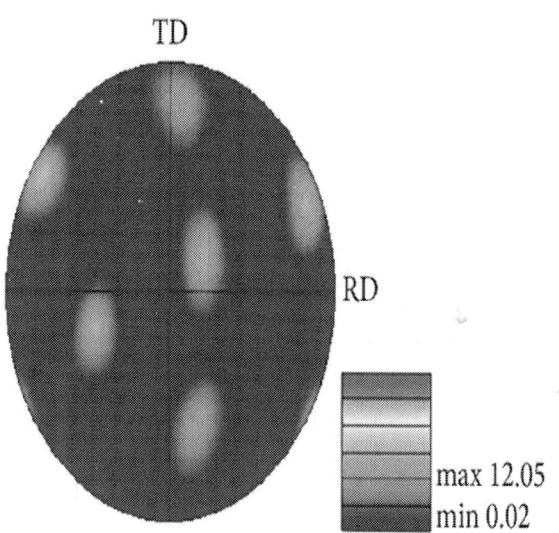

max 12.05
min 0.02

(c)

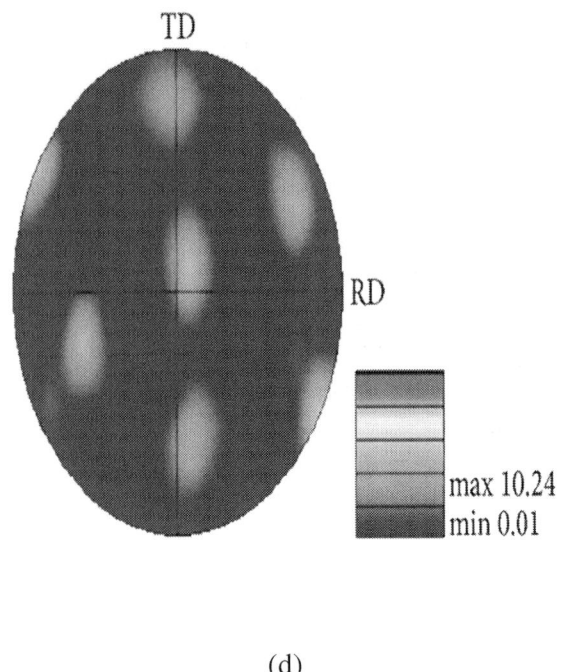

(d)

Figure 6: Pole figures of the stir zone in (a) SPCC/SPCC, (b) SPCC/SUS, (c) SUS/SPCC, and (d) SUS/SUS, respectively; RD and TD correspond to rolling and transversal directions, respectively.

Hardness profiles were made with a spacing of 0.3 mm along three parallel lines: two lines in the upper sheet and one line in the lower sheet (Figure 7(a)); these profiles were collected for FSSW joints of each of the four material combinations tested (Figures 7(b)–7(e)). The hardness profiles in Figures 7(b)–7(e) show typical hardness distributions across the base metal, the heat affected zone (HAZ), and the stir zone. The hardness of the stir zone is generally higher than that of the base metal for both SPCC and SUS due to the large plastic deformation and fine-grained microstructure in the stir zone [22]; a slight decrease of the hardness in the HAZ is also observed.

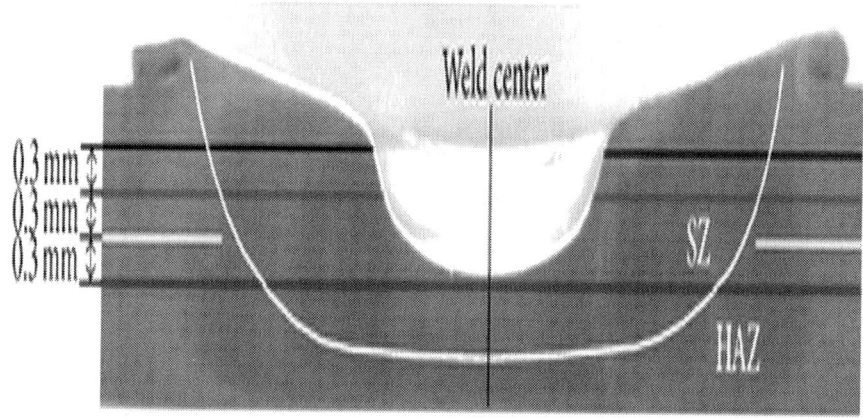

(a)

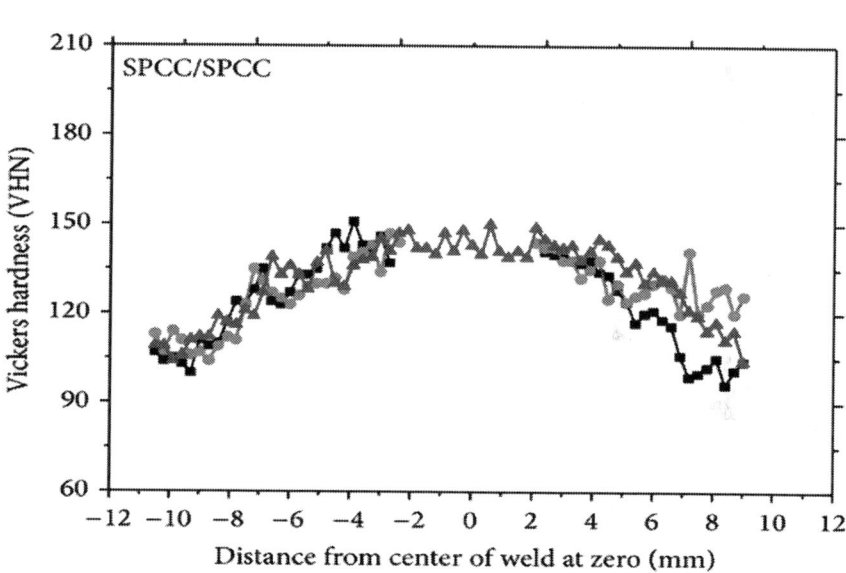

(b)

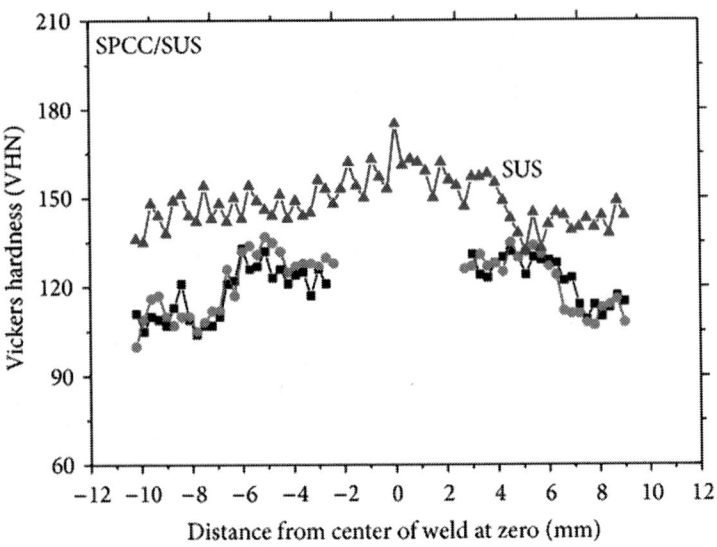

(c)

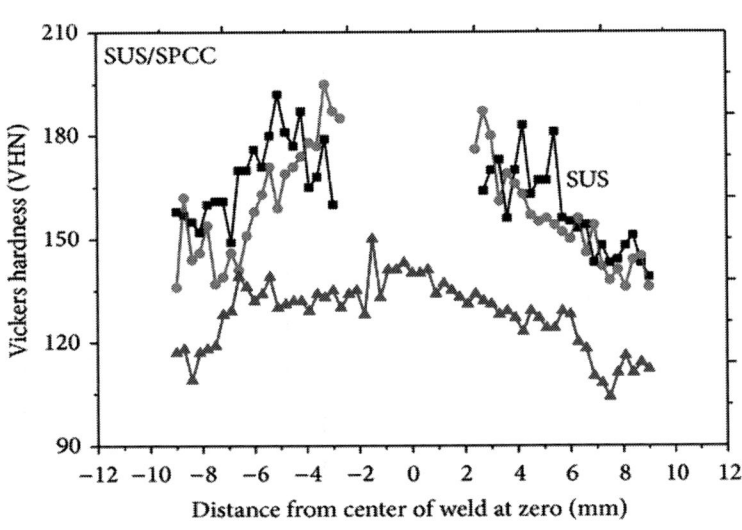

(d)

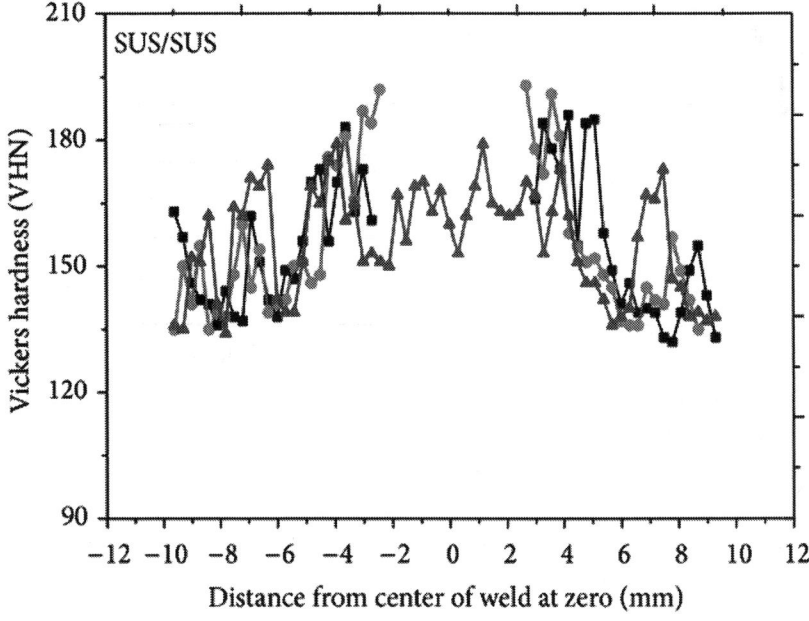

(e)

Figure 7: (a) Locations of the three parallel hardness traverses for a SPCC/SPCC joint; hardness profiles of FSSW joint cross sections, showing typical hardness distributions across the base metal, the HAZ, and the stir zone: (b) SPCC/SPCC, (c) SPCC/SUS, (d) SUS/SPCC, and (e) SUS/SUS.

Mechanical Behavior under Combined Loads

Load-displacement curves for FSSW joints of four different material combinations at four different loading angles clearly show that the maximum failure load and toughness of the joints depend on the material combinations and loading angles (Figure 8); the maximum load decreases as the loading angle increases for all material combinations studied as listed in Table 3. This tendency of the maximum load is similar to that observed in RSW experiments reported by Lin et al. [18] and Song and Huh [20].

Table 3: The quasi-static failure loads of the FSSW joints under various loading conditions

Loading angle (°)	Maximum load* (kN)			
	SPCC/SPCC	SPCC/SUS	SUS/SPCC	SUS/SUS
0	7.81 (0.095)	8.42 (0.060)	8.60 (0.230)	10.00 (0.201)
15	7.70 (0.032)	7.75 (0.008)	6.54 (0.183)	8.81 (0.283)
22	7.40 (0.077)	7.60 (0.062)	6.43 (0.220)	8.43 (0.092)
30	6.93 (0.060)	7.40 (0.073)	6.10 (0.090)	8.30 (0.052)

Average of the results of two FSSW specimens; values in the parentheses are the standard deviations.

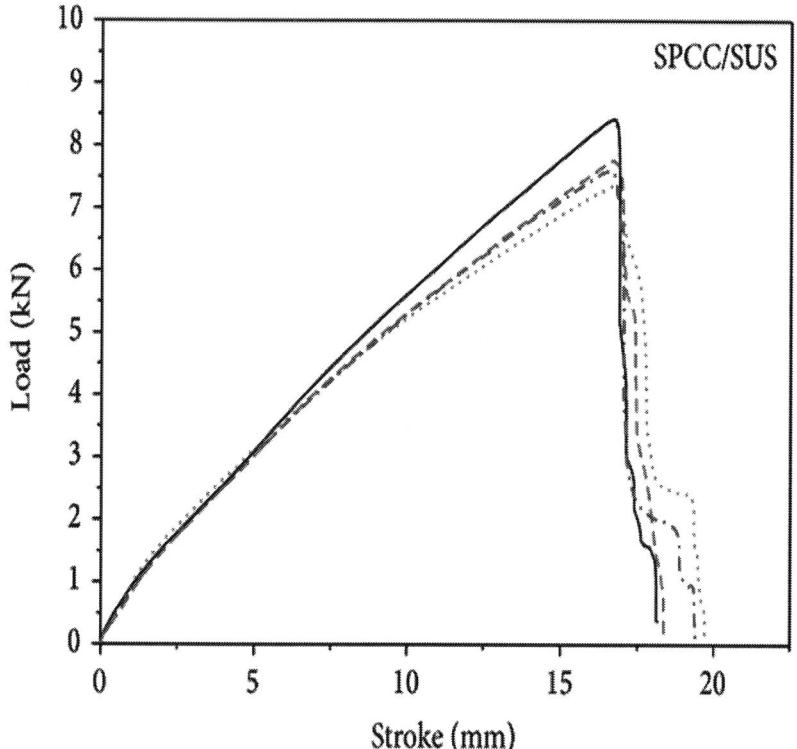

(a)

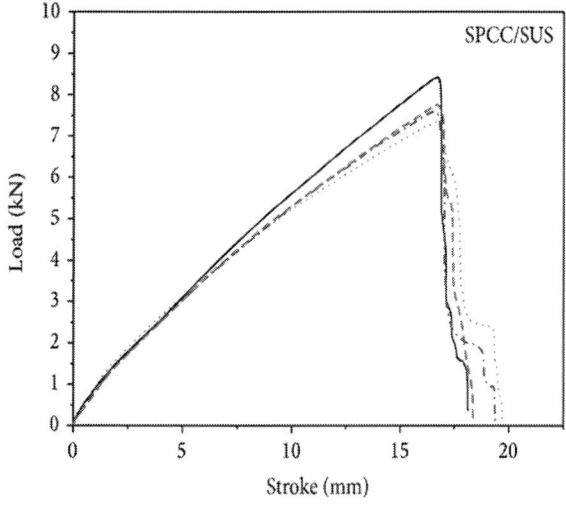

(b)

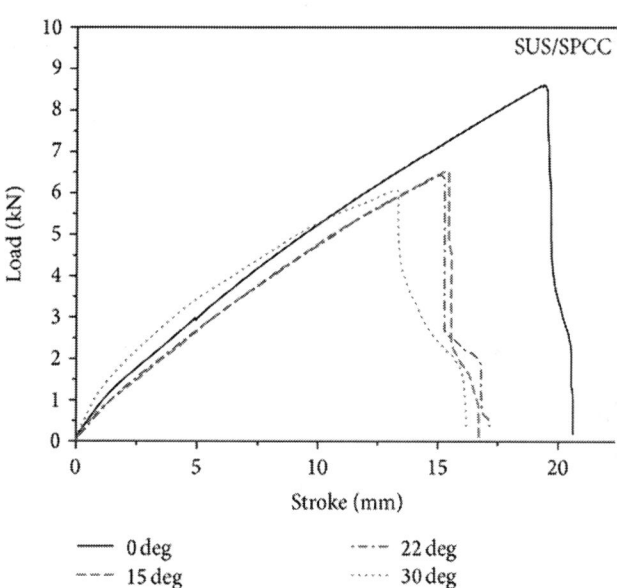

(c)

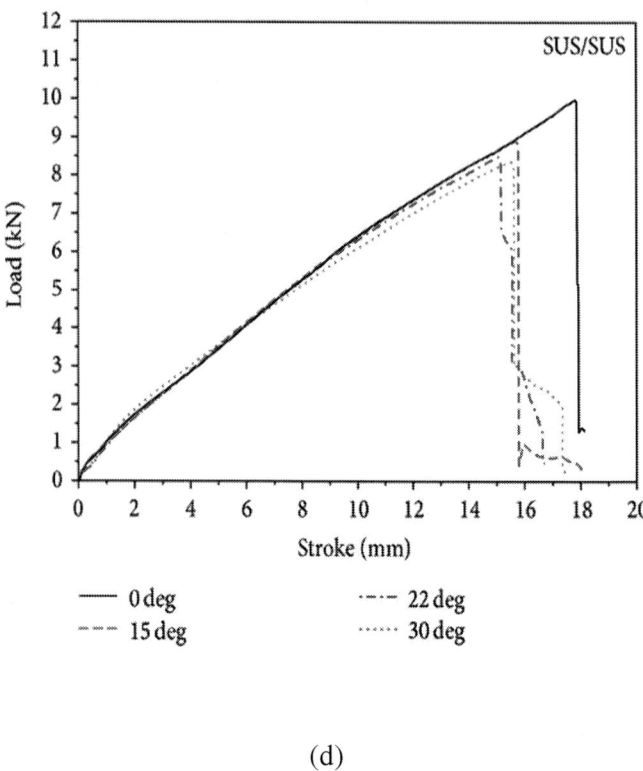

(d)

Figure 8: Load-displacement curves for FSSW joints of (a) SPCC/SPCC, (b) SPCC/SUS, (c) SUS/SPCC, and (d) SUS/SUS at four different loading angles.

Under a pure opening load ($\phi = 0°$), failure occurred by a typical nugget pullout mode. By contrast, under a combined load, the joint›s failure initiated with a nugget rotation due to the shear component of the combined load, followed by complete separation of the upper sheet by tearing off from the joint. A comparison of the top and bottom views of the completely separated SPCC/SPCC FSSW joints under a pure opening load ($\phi = 0°$, Figures 9(a) and 9(b)) and a combined load of $\phi = 30°$ (Figures 9(c) and 9(d)) shows that nugget rotation occurred under the combined load (Figures 9(c) and 9(d)), while a typical nugget pullout fracture occurred under the pure opening load (Figures 9(a) and 9(b)). As shown in Figure 9(d), the upper sheet was torn off at the final stage of the failure as the rupture propagated along the circumference of the nugget. A similar failure mechanism was observed for RSW by Lin et al. [18], Song and Huh [20], and Song et al. [23].

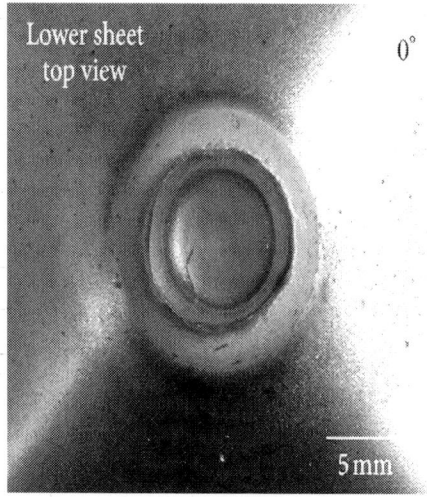

(a)

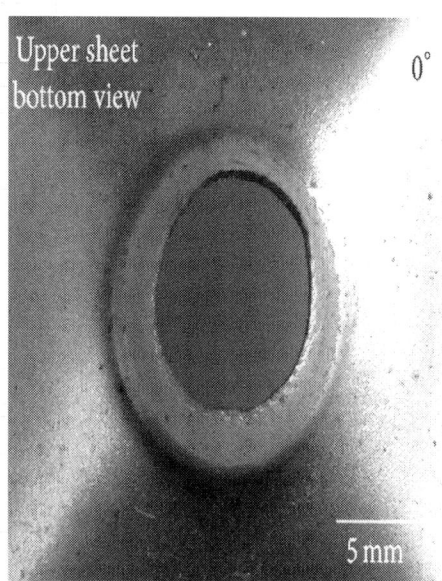

(b)

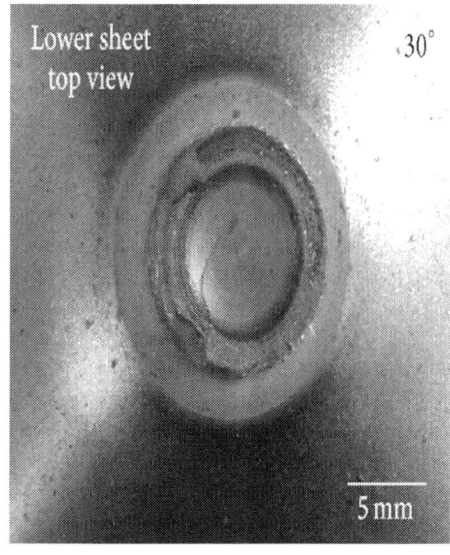

(c)

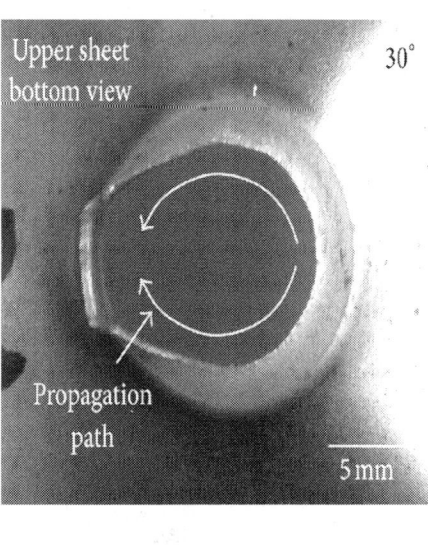

(d)

Figure 9: Top and bottom views of the completely separated SPCC/ SPCC FSSW joints under ((a), (b)) a pure opening load ($\phi = 0°$) and ((c), (d)) a combined load of $\phi = 30°$, respectively.

The cross-sectional optical micrograph of a completely failed SPCC/SPCC FSSW joint under a pure opening load ($\phi = 0°$) shows that the typical nugget pullout failure occurred by uniform necking/shear along the circumferential boundary of the nugget as shown in Figure 10(a). The cross-sectional optical micrograph of a completely failed FSSW joint under a combined load of ($\phi = 22°$) (Figure 10(b)) shows that the failure of the joint under a combined load was initiated by necking and shear at the stretching side with respect to the shear component of the load (marked as A in Figure 10(b)), even though signs of necking are also observed on the opposite side. The rupture then propagated along the circumference of the weld nugget. Finally, the upper sheet was completely torn off from the lower sheet at nearly the opposite side from the location where the rupture initiated (marked as B in Figure 10(b)). The failures of FSSW joints for the material combinations of SPCC/SUS, SUS/SPCC, and SUS/SUS and at different combined loading angles $\phi = 15°$ and $30°$ were quite similar to the result shown in Figure 10 and not shown here.

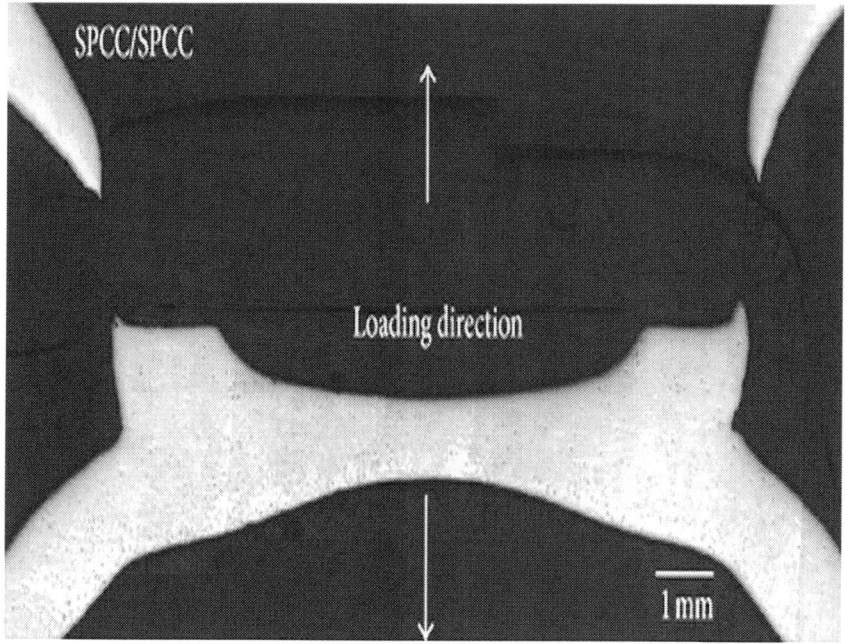

(a)

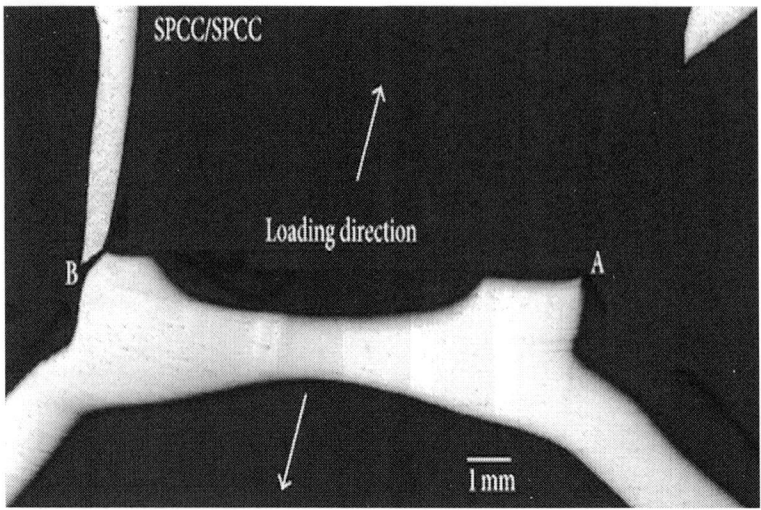

(b)

Figure 10: Cross-sectional macrographs of completely failed SPCC/SPCC FSSW joints under (a) a pure opening load ($\phi = 0°$) and (b) a combined load of $\phi = 22°$, respectively.

The maximum loads under combined loads were simply decomposed into the axial and shear components using (1) for the four different material combinations. A failure contour then can be constructed in terms of the axial and shear load. Several failure criteria have been proposed to describe the failure of RSW under combined loads. Lee et al. [17] proposed a failure criterion based on the normal and shear failure loads of the weld, which were determined under pure opening and shear loads, respectively. Under combined loads, they proposed the following failure criterion:

$$\left(\frac{F_n}{F_N} \right)^n + \left(\frac{F_s}{F_S} \right)^n = 1,$$

(2)

Where F_n and F_s are the applied normal and shear loads, respectively, F_N and F_S are the normal and shear failure loads of the spot weld, respectively, and n is the fitting parameter. In general, n is set to 2 to fit the experimental results.

Lin et al. [18] proposed an engineering failure criterion in terms of the axial and shear loads with consideration of the sheet thickness and the nugget radius under combined loads. Their failure criterion, based on the lower bound analysis under combined opening and shear loads, is expressed as

$$\left[1 - 2\alpha + 2\alpha^2\right]\left(\frac{F_n}{2\pi r t \tau_0}\right)^2$$

$$+ \left[\frac{1}{3} + \left(\frac{4t}{2\pi r}\right)\right]\left(k\frac{F_s}{2\pi r t \tau_0}\right)^2 = 1,$$

(3)

where τ_0 is the shear yield strength, α is the loading parameter, and k is the fitting constant. For a combined load, $= 0.5$ is used for square-cup specimens [18].

Recently, Song and Huh [20] also proposed a failure criterion to describe the failure behavior of RSW joints under combined loads:

$$\left(\frac{F_n}{F_N}\right)^2 + \beta\left(\frac{F_n}{F_N}\right)\left(\frac{F_s}{F_S}\right) + \left(\frac{F_s}{F_S}\right)^2 = 1.$$

(4)

Here, again, F_N and F_S are the normal and shear failure loads of the spot weld, respectively. The variable β is the failure parameter that can be obtained by least-squares fitting to minimize the discrepancy between the experimental result and the interpolated one. The shape of the failure curve is elliptic when $\beta = 0$, and in this condition is identical to the failure criterion proposed by Lee et al. [17].

Based on the experimental results and above mentioned failure criteria, failure contours for FSSW joints under combined loads were constructed in terms of the axial and shear loads for each of the four different material combinations (Figure 11). Note that in constructing the failure contours, the experimental result of lap-shear specimens [21] was used as the shear failure load F_s. Also note that for the failure contours of dissimilar FSSW joints based on (3), the shear yield strength of the upper sheet material was used since the fracture mainly occurred in the upper sheet material. Comparison of the failure contours suggests that the failure criterion proposed by Lee et al. [17] is inadequate to describe the failure of FSSW joints under opening-dominant combined loads. Although the experimental result of FSSW joints with SPCC on the upper sheet (Figures11(a)-11(b)) is relatively consistent with the failure criterion suggested by Lin et al. [18], the experimental result of FSSW joints with SUS on the upper sheet (Figures 11(c)-11(d)) does not agree well with that criterion. On the other hand, the experimental results for all four different material combinations can agree well with the failure criterion proposed by Song and Huh [20] by selecting different values of β. It is interesting to note that the value of β strongly depends on the material on the upper sheet of the joint; the FSSW joints with a softer and more ductile material on the upper sheet (SPCC/SPCC and SPCC/SUS) have considerably lower values of β. Actually, comparison of the failure contours based on Song and Huh [20] in terms of the normalized axial and shear loads, which were normalized by the corresponding normal (pure opening, $\phi = 0°$) and shear $\phi = 90°$ failure loads, respectively, confirms the dependence of the failure contours on the upper sheet material (Figure 12). The shape of the failure contours is close to elliptic with SPCC as the upper sheet, while the failure contours take the form of a relatively straight line with SUS as the upper sheet.

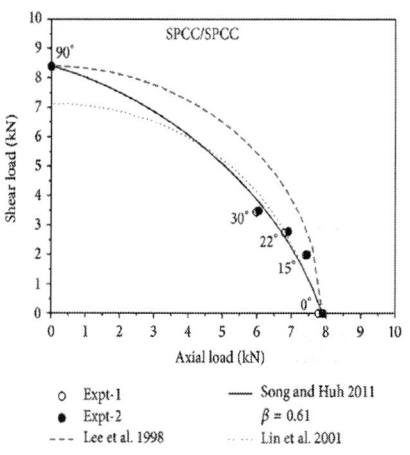

(a)

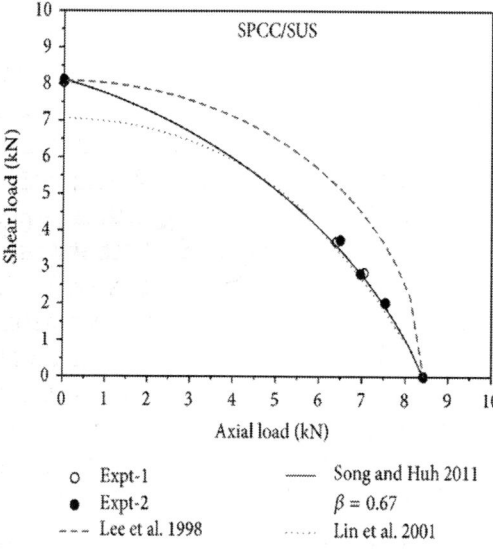

(b)

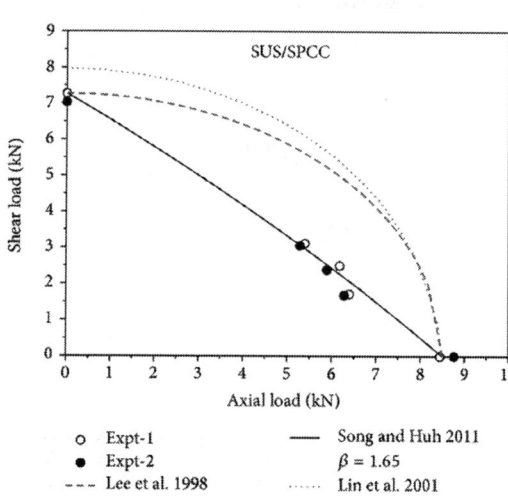

(c)

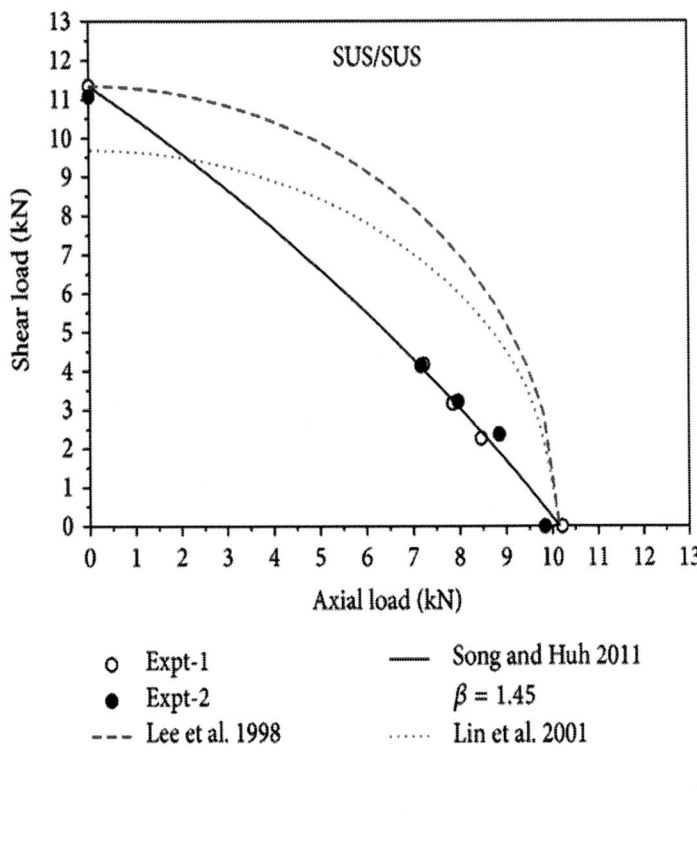

(d)

Figure 11: Comparison of the experimental result with conventional failure criteria: (a) SPCC/SPCC, (b) SPCC/SUS, (c) SUS/SPCC, and (d) SUS/SUS.

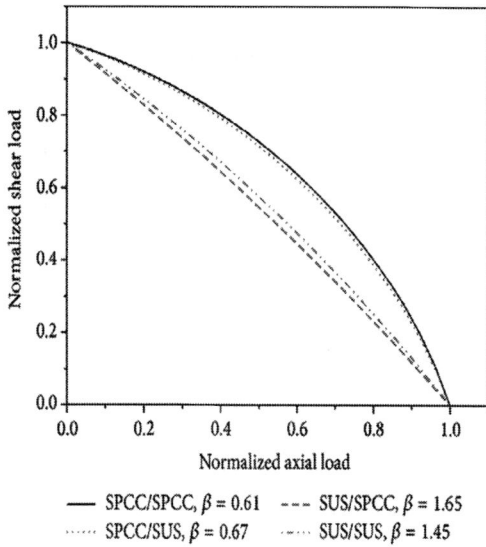

Figure 12: Normalized failure contours for the FSSW joints with four different material combinations based on the failure criterion of Song and Huh [20].

CONCLUSIONS

Mechanical behaviors of FSSW joints of two dissimilar ferrous alloys under opening-dominant combined loads were experimentally investigated. Defect-free spot joints were successfully fabricated with four different material combinations. EBSD analysis shows that extremely fine homogeneous grains developed in the stir zone, while the texture of dissimilar FSSW joints depends on the upper sheet material. The failure contours for the FSSW joints under combined loads were constructed in terms of the axial load and shear load by modifying existing failure criteria for RSW. The shape of the failure contour also depends on the upper sheet material. The failure contours are nearly elliptic in shape when the upper sheet is SPCC and are relatively straight lines when the upper sheet is SUS. The results of the present study also suggest that the mechanical and material properties of FSSW joints of dissimilar ferrous alloys are improved when the lap joint is designed with the "harder" material on the bottom and the "softer" material on top.

ACKNOWLEDGMENTS

This research was financially supported by the Ministry of Education (MOE) and National Research Foundation of Korea (NRF) through the Human Resource Training Project for Regional Innovation. Michael Miles acknowledges support from National Science Foundation Grant CMMI-1131203. H.-H. Cho and H. N. Han were supported by Basic Science Research Program through the National Research Foundation of Korea (NRF) and funded by the Ministry of Science, ICT, and Future Planning (2013008806).

REFERENCES

1. W. M. Thomas, E. D. Nicholas, J. C. Needham, P. Temple-smith, S. W. K. W. Kallee, and C. J. Dawes, "Friction stir welding," UK Patent Application GB 2 306 366 A, 1991.

2. W. Yuan, R. S. Mishra, S. Webb et al., "Effect of tool design and process parameters on properties of Al alloy 6016 friction stir spot welds," Journal of Materials Processing Technology, vol. 211, no. 6, pp. 972–977, 2011. · ·

3. D.-A. Wang and S.-C. Lee, "Microstructures and failure mechanisms of friction stir spot welds of aluminum 6061-T6 sheets," Journal of Materials Processing Technology, vol. 186, no. 1-3, pp. 291–297, 2007. · ·

4. E. Folkhard, Welding Metallurgy of Stainless Steels, spring, New York, NY, USA, 1988.

5. O. P. Khanna, Welding Technology, Dhanpat Rai Publications, New Delhi, India, 16th edition, 2007.

6. Z. Feng, M. L. Santella, S. A. David, et al., "Friction Stir Spot Welding of Advanced High-Strength Steels -A Feasibility Study, SAE World Congress, Detroit, Mich, USA, 2005.

7. S. W. Baek, D.-H. Choi, C.-Y. Lee et al., "Structure-properties relations in friction stir spot welded low carbon steel sheets for light weight automobile body," Materials Transactions, vol. 51, no. 2, pp. 399–403, 2010. · ·

8. Y. Hovanski, M. L. Santella, and G. J. Grant, "Friction stir spot welding of hot-stamped boron steel,"Scripta Materialia, vol. 57, no. 9, pp. 873–876, 2007. · ·

9. M. P. Miles, C. S. Ridges, Y. Hovanski, J. Peterson, M. L. Santella, and R. Steel, "Impact of tool wear on joint strength in friction stir spot welding of DP 980 steel," Science and Technology of Welding and Joining, vol. 16, no. 7, pp. 642–647, 2011. · ·

10. Y. F. Sun, H. Fujii, N. Takaki, and Y. Okitsu, "Microstructure and mechanical properties of mild steel joints prepared by a flat friction stir spot welding technique," Materials and Design, vol. 37, pp. 384–392, 2012. · ·

11. Y. J. Chao, "Ultimate strength and failure mechanism of resistance spot weld subjected to tensile, shear, or combined tensile/ shear loads," Journal of Engineering Materials and Technology, Transactions of the ASME, vol. 125, no. 2, pp. 125–132, 2003. ·

12. P. Wung, "A forced-based failure criterion for spot weld design," Experimental Mechanics, vol. 41, no. 1, pp. 107–113, 2001.

13. P. Wung, T. Walsh, A. Ourchane, W. Stewart, and M. Jie, "Failure of spot welds under in-plane static loading," Experimental Mechanics, vol. 41, no. 1, pp. 100–106, 2001.

14. D. Radaj, "Stress singularity, notch stress and structural stress at spot-welded joints," Engineering Fracture Mechanics, vol. 34, no. 2, pp. 495–506, 1989.

15. D. Radaj and S. Zhang, "On the relations between notch stress and crack stress intensity in plane shear and mixed mode loading," Engineering Fracture Mechanics, vol. 44, no. 5, pp. 691–704, 1993.

16. S. Zhang, "Approximate stress formulas for a multiaxial spot weld specimen," Welding Journal, vol. 80, no. 8, pp. 201s–203s, 2001.

17. Y.-L. Lee, T. J. Wehner, M.-W. Lu, T. W. Morrissett, and E. Pakalnins, "Ultimate strength of resistance spot welds subjected to combined tension and shear," Journal of Testing and Evaluation, vol. 26, no. 3, pp. 213–219, 1998.

18. S.-H. Lin, J. Pan, S.-R. Wu, T. Tyan, and P. Wung, "Failure loads of spot welds under combined opening and shear static loading conditions," International Journal of Solids and Structures, vol. 39, no. 1, pp. 19–39, 2001. · ·

19. S.-H. Lin, J. Pan, T. Tyan, and P. Prasad, "A general failure criterion for spot welds under combined loading conditions," International Journal of Solids and Structures, vol. 40, no. 21, pp. 5539–5564, 2003. · ·

20. J. H. Song and H. Huh, "Failure characterization of spot welds under combined axialshear loading conditions," International Journal of Mechanical Sciences, vol. 53, no. 7, pp. 513–525, 2011. · ·

21. M. A. M. Hossain, M. T. Hasan, S. T. Hong, M. Miles, H. H. Cho, and H. N. Han, "Failure behaviors of friction stir spot welded joints of dissimilar ferrous alloys under quasi-static shear loads," International Journal of Materials and Product Technology, vol. 48, no. 1–4, pp. 179–193, 2014.

22. Y. Miyano, H. Fujii, Y. Sun, Y. Katada, S. Kuroda, and O. Kamiya, "Mechanical properties of friction stir butt welds of high nitrogen-containing austenitic stainless steel," Materials Science and Engineering A, vol. 528, no. 6, pp. 2917–2921, 2011.

23. J. H. Song, H. Huh, J. H. Lim, and S. H. Park, "Effect of tensile speed on the failure load of a spot weld under combined loading conditions," International Journal of Modern Physics B, vol. 22, no. 9–11, pp. 1469–1474, 2008.

Investigations on Mechanical and Tribological Behaviour of Particulate Filled Glass Fabric Reinforced Epoxy Composites

Bhadrabasol Revappa Raju[1], Bheemappa Suresha[2], Ragera Parameshwarappa Swamy[3,] and Bannangadi Swamy Gowda Kanthraju[4]

[1]Department of Mechanical Engineering, PES Institute of Technology and Management, Shivamogga, India

[2]Department of Mechanical Engineering, The National Institute of Engineering, Mysore, India

[3]Department of Mechanical Engineering, University B.D.T. College of Engineering, Davangere, India

[4]Department of Mechanical Engineering, Don Bosco Institute of Technology, Bangalore, India

ABSTRACT

The aim of the research article is to study the mechanical and two-body abrasive wear behaviour of alumina (Al_2O_3) filled glass fabric reinforced epoxy (G-E) composites. Alumina filled G-E composites containing 0, 5, 7.5 and 10 wt% were prepared using the hand lay-up technique followed by compression molding. The mechanical properties such as tensile strength, hardness and tensile modulus were investigated in accordance with ASTM standards. Two-body abrasive wear studies were carried out using a pin-on-disc wear tester under multi-pass condition against the water proof silicon carbide abrasive paper. From the experimental investigation, it was found that the presence of Al_2O_3 filler improved the tensile strength and tensile modulus of the G-E composite. Inclusion of Al_2O_3 filler reduced the specific wear rate of G-E composite. The results show that in abrasion mode, as the filler loading increases the wear volume decreases and increased with increasing abrading distance. The excellent wear resistance was obtained for Al_2O_3filled G-E composites. Furthermore, 10 wt% filler loading gave a very less wear loss. Finally, the scanning electron microscopic observations on the wear mechanisms Al_2O_3filled G-E composites were discussed.

INTRODUCTION

Polymer based materials are finding increasing use in many applications owing to their strength, lightness, ease of processing and availability of wider choice of systems [1]. Polymer and their composites are finding ever increasing usage for numerous industrial applications such as bearing material, rollers, seals, gears, cams, wheels, clutches and transmission belts etc. [2-5]. The importance of tribological properties convinced many researchers to study the wear behaviour and to improve the wear resistance of polymeric composites. For fiber reinforced polymer matrix composites, the process of material removal in the abrasive wear process involves four different mechanisms microploughing, microcutting, microfatigue and microcracking [6].

Wear is defined as damage to a solid surface, generally involving progressive loss of material, due to relative motion between that surface and contacting substance or substances [7]. Abrasive wear is the most important among all the forms of wear because it contributes almost 64% of the total cost of wear [8]. Abrasive wear is caused due to hard particles or hard protuberances that are forced against and move along a solid surface [9]. In two-body abrasion, wear is caused by hard protuberances on one surface which can only slide over the other. Tribo-engineering materials and are invariably used in mechanical components, where wear performance in nonlubricated condition is a key parameter in the material selection [10, 11]. Carbon, graphite, glass and aramid fabrics are the most commonly used fabrics for fiber reinforced polymer composites especially for making tribocomponents and aircraft structures that encounter harsh operating conditions such as high stresses, speeds, temperatures, etc. [12-14].

However, the woven fabric composites are getting acceptance in many engineering applications such as in circuit board, marine, aerospace, transportation and other industries for several reasons. They are commonly used in industry to manufacture composite components due to their ease of use, improve structural performance and reduction in cost. They provide better resistance to impact than unidirectional composites and display behavior that is closer to that of a fully isotropic material [14-16]. Modification of woven fabric reinforced composites by incorporation of fillers has been a popular research activity in the plastics industry since the properties of the resultant materials may be significantly changed by the introduction of fillers and fabrics [17].

A literature survey indicated that the short fiber reinforcement, in general, led to the deterioration in the abrasive wear resistance of the matrix [18]. Fabric reinforcement, on the other hand, improved the abrasion resistance of the polymers [19]. Many researchers studied the two-body wear behaviour of polymers in general and polymer composites in particular [20-25]. In some of the literature concerning abrasive wear of polymers, Friedrich [26] investigated the abrasive wear behaviour of the epoxy reinforced with carbon, glass and aramid fabrics and reported the wear performance of the fabrics in the order Aramid > glass > carbon. Bijwe et al., [28] tested polyamide 6, polytetrafluoroethylene (PTFE) and their various composites in abrasive wear under dry and multi-pass conditions against silicon carbide (SiC) paper on pin-on-disc arrangement. Suresha et al. [29] investigated the friction and wear behavior of glass-epoxy composite with and without graphite. They fabricated neat glass-epoxy composite and graphite filled glass-epoxy composite with three different percentages of filler.

They concluded the graphite filled glass epoxy composite shows higher resistance to sliding wear as compared to plain glass-epoxy composites. To evaluate the possibility of improving the mechanical and abrasive wear of glass fabric reinforced epoxy composites and elucidate the abrasive wear mechanisms, In view of the above, this research article reports a study on mechanical and two-body abra- sive wear performance of unfilled and Al_2O_3 filled G-E composites.

EXPERIMENTAL DETAILS

Materials and Fabrication

The matrix material system selected is an Epoxy resin (LAPOX L-12 with density 1.16 g/cm^3) supplied by ATUL India Ltd., Gujarat, India. Woven glass plain weave fabrics made of 360 g/m^2; containing E-glass fibers of diameter of about 12 μm have been used as the reinforcing material in all the composites. The fillers chosen were aluminum oxide (Al_2O_3). The average particle size of Al_2O_3 micro particles is about 10 μm size. The details of the constituents selected for the present work are listed in Table 1. As regards to the processing, on a Teflon sheet, E-glass woven fabric was placed over which the epoxy matrix system consisting of epoxy and hardener was smeared. Dry hand lay-up technique was employed to fabricate the composites. The stacking procedure consists of placing the fabric one above the other with the resin mix well spread between the fabrics. A porous Teflon film was again used to complete the stack. To ensure uniform thickness of the sample, a 3 mm spacer was used. The mould plates were coated with release agent in order to aid the ease of separation on curing. The cast of each composite after 12 h of impregnation and dried for 2 h at 100°C followed by compression molding at a temperature of 390°C and a pressure of 7.35 MPa. The slabs so prepared measured 250 mm × 250 mm × 3 mm in size. To prepare different wt% of Al_2O_3 filled G-E composites, besides the epoxy hardener mixture, additional wt% of Al_2O_3 particles were included to form the resin mix. The details of the composites selected for the present work are listed in **Table 2**. The percentage of the glass fiber in the composite is 60 by wt%. Mechanical and abrasive wear test samples were prepared according to ASTM standard from the cured laminates using a diamond tipped cutter.

Physico-Mechanical Tests

The density of the composites was determined by using a high precision electronic balance (Mettler Toledo, Model AX 205) using the Archimedes principle. Hardness (Shore-D) of the samples was measured as per ASTM D2240, by using a Hiroshima make hardness tester (Durometer). Five readings at different locations were noted and average value is reported. Tensile properties were measured using a Universal testing machine in accordance with the ASTM D-3039 procedure at a cross head speed of 5 mm/min and a gauge length of 50 mm. The tensile strength and modulus were determined from the stress-strain curves. Five samples were tested in each set and the average value was reported. The tensile test was carried out on a fully automated Lloyd LR-20 kN Universal testing machine connected to a computer with DAPMAT software.

Table 1: Physical and mechanical properties of the constituents selected for the present work

Property	Epoxy	Glass fibers	Al203 filler
Density (g.:cm³)	1.16	2.54	3.89
Tensile strength (MPa)	110	3400	260 - 300
Tensile modulus (GPa)	4.1	72.3	375

Table 2: Composites selected for the present study

Sample name (designation)	Glass fiber (wt %)	Epoxy (wt %)	Al_2O_3 filler (wt %)
Glass fabric reinforced epoxy(G-E)	60	40	

Aluminium oxide filled G-E(5% Al_2O_3-G-E)	60	35	5
Aluminium oxide filled G-E(7.5% Al_2O_3-G-E)	60	32.5	7.5
Aluminium oxide filled G-E(10% Al_2O_3-G-E)	60	30	10

Two-Body Abrasive Wear Test

Two-body abrasive wear tests were performed using a Pin-on-Disc machine according to ASTM G99 standards. Test samples were prepared after proper cutting and polishing to 6 mm × 6 mm × 3 mm size. The composite sample was abraded against the water proof silicon carbide (SiC) abrasive papers of 320 and 600 grit size at a constant running speed of 175 rpm in multi-pass condition Figures 1(a) and (b). During wear test, the sample is so placed in such a way that the fibers are parallel and anti-parallel with respect to the abrading direction and the abrading plane. A constant normal load of 10 N was applied. The weight loss measurements were carried out for four abrading distance of 7.5, 15, 22.5 and 30 m. Before and after wear testing, samples were cleaned with brush to remove wear debris. The wear was measured by the loss in weight (Mettler: TOLEDO, 0.1 mg accuracy), which was then converted into wear volume using the measured density data.

The specific wear rate (K_s) was calculated from the equation:

$$K_s = \frac{\Delta V}{L \times D} \, \text{m}^3 / \text{Nm}$$

(1)

Where ΔV is the volume loss in m^3, L is the load in Newtons and D is the abrading distance in meters.

After wear test, the worn surfaces of specimens were examined using a scanning electron microscope (JSM 840A model and JEOL make). Before the examinations, a thin gold film was coated on the worn surface by sputtering to achieve a conducting layer.

RESULTS AND DISCUSSION

Effect of Filler Loading on Density

The measured densities of the samples are listed in Table 3. Comparing the results it was observed that the inclusion of ceramic filler into G-E showed higher density. The density of 10 wt% Al_2O_3 filled G-E is 2.3 which are higher when compared to other composites. This is because of the filler Al_2O_3 has a higher density. The densities of all micro particles filled G-E is higher than the density of unfilled G-E composites.

(a)

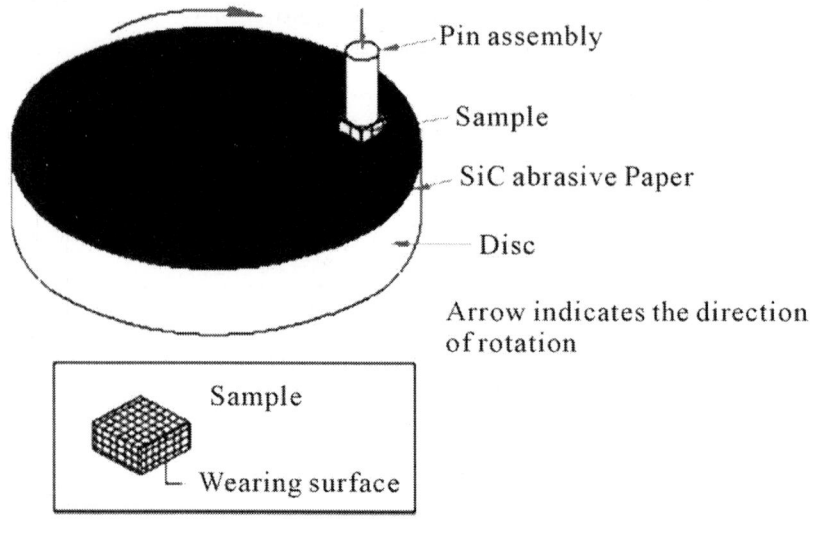

(b)

Figure 1: (a) Pin-on-disc wear test apparatus; (b) Rotating disc with SiC paper and composite sample.

Table 3: Physico-mechanical properties of G-E and Al_2O_3 filled G-E composites

Sample code	G-E	5% Al_2O_3-G-E	7.5% Al_2O_3-G-E	10% Al_2O_3-G-E
Density, (g/cm³)	1.984	2.12	2.23	2.3
Hardness (Shore-D)	63	66	69	72
Tensile strength, σ (MPa)	254	324	343	352
Tensile modulus, E (GPa)	8.34	10.6	11.26	11.55
Elongation, e (mm)	7.1	6.4	6.2	5.9

Effect of Filler Loading on Hardness

The hardness of G-E composite increased with increase of micro particle filler loading. By using the Duro-hardness tester, the hardness of the composites is measured; the values recorded are given in Table 3, it can be seen that the Al_2O_3 filler greatly increased the hardness of G-E, which can be attributed to the higher hardness and more uniform dispersion of Al_2O_3 filler. The higher hardness is exhibited by the 10 wt% Al_2O_3 filled G-E compared to other composites. The hardness of 10 wt% Al_2O_3 filled G-E composite is 72, which is highest among all the composites tested. Particulate filled G-E composites with sufficient surface hardness are resistant to in-service scratches that can compromise fatigue strength and lead to premature failure. Therefore, under an indentation loading, micro particles would undergo elastic rather than plastic deformation, as compared to unfilled G-E composites. The improvement in hardness with the incorporation of filler can be explained as follows: under the action of a compressive force, the thermoset matrix phase and the solid fiber and filler phase will be pressed together, touch each other and offer resistance. Thus the interface can transfer load more effectively although the interfacial bond may be poor. This results in enhancement of hardness of Al_2O_3 filled G-E composites.

Tensile Properties

The typical load-deformation curves of unfilled and particulate filled G-E campsites are shown in Figure 2 and the measured mechanical test results are listed in Table 3. The average ultimate tensile strength values for G-E composites with 0, 5, 7.5 and 10 wt% of Al_2O_3 filler are 254, 324, 343 and 352 MPa, respectively. The tensile strength of the Al_2O_3 filled G-E composites increased with increasing Al_2O_3 up to 7.5 wt%, because of the uniform dispersion of Al_2O_3 filler in G-E. However, the in- crease in tensile strength is marginal beyond 7.5 wt% of Al_2O_3 filler loading. This could be attributed the uniform dispersion of Al_2O_3 filler in G-E. The surface modified Al_2O_3 can interact with the fiber surface and hydrogen bonding increases and leads to the better interaction with glass fiber and epoxy. Addition of ceramic fillers increases the effective mechanical interlocking, which in turn increases the frictional force between the fiber and matrix. It can be seen from **Table 3** that the tensile modulus of Al_2O_3 filled G-E composites increases as the wt. fraction of the filler increases. Again there is a reduction in the elongation at break of the composites with an increase in the weight fraction of the filler. This is due to the fact that the Al_2O_3 filler is hard and also highly brittle. As the wt. fraction of Al_2O_3 filler increase, the tensile modulus of the G-E composites increases, but at the same time the system becomes more brittle. The increase in the tensile strength with wt. fraction of filler is attributed to the high modulus of ceramic filler which are dispersed uniformly in the fabric layers of G-E composites. Adding Al_2O_3 did not alter the tensile modulus appreciably except at 5 wt% filler loading. The average Young's modulus values for composites with 0, 5, 7.5 and 10 wt% Al_2O_3 are 8.34, 10.6, 11.26 and 11.55 GPa, respectively. Young's modulus is mainly dependent on the matrix deformation of the composite and increases as the slope of the load-deformation curve at the initial stage and is practically not much influenced by the interfacial strength between fiber and matrix. Generally, the addition of ceramic fillers and glass fiber reduces the elongation at break because of the lower elongation at break values of ceramic fillers and glass fiber compared to that of epoxy matrix. Also the effects of filler loading on the mechanical properties of particulate filled G-E composites were studied and it can readily be seen from the data given in **Table 3** and Figures 1(a) and (b).

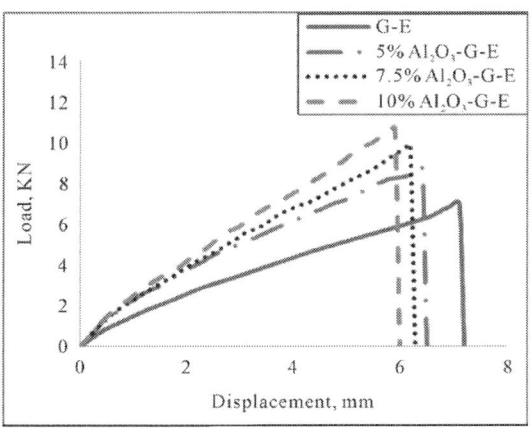

Figure 2: Typical load v/s displacement curves of G-E & Al_2O_3 filled G-E samples.

At the filler loading 0 - 10 wt% Comparing the results, it can be seen that Al_2O_3 filled G-E samples show improved mechanical properties, confirming the effect of Al_2O_3 filler inclusion. The addition of Al_2O_3 particles causes a dispersion of these particles in the matrix which impede to the propagation of failure along the loading direction. Thus the failure would propagate easily in those directions where the dispersed concentration is less leading to increased tensile strength, tensile modulus, and lower elongation.

Wear Volume

The variation in abrasive wear volume of composites worn in 320 and 600 grit SiC paper at 10 N against abrading distance under multi-pass condition is shown in Figures 3(a) and (b) respectively The wear data of the composites reveal that the wear volume tends to increase near linearly with increasing abrading distance and strongly depends on the grit size of the abrasive paper. From Figures 3(a) and (b), it is obvious that the wear volume of composites worn on two different SiC papers increased with increasing abrading distance. Wear volume of unfilled G-E is much higher than those of filled G-E composites and also the wear volume decreased with the increasing weight percentage of filler. In addition, the highest wear volume is obtained from specimens worn on 320 grit SiC paper Figure 3(a).

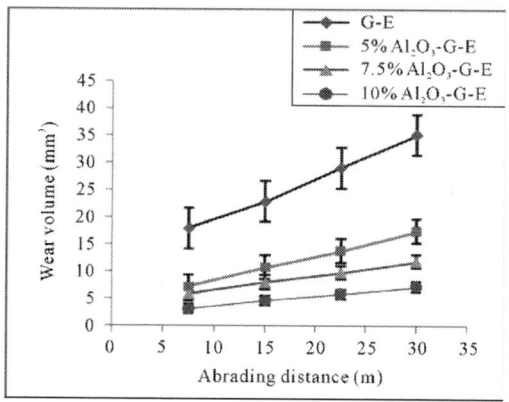

(a)

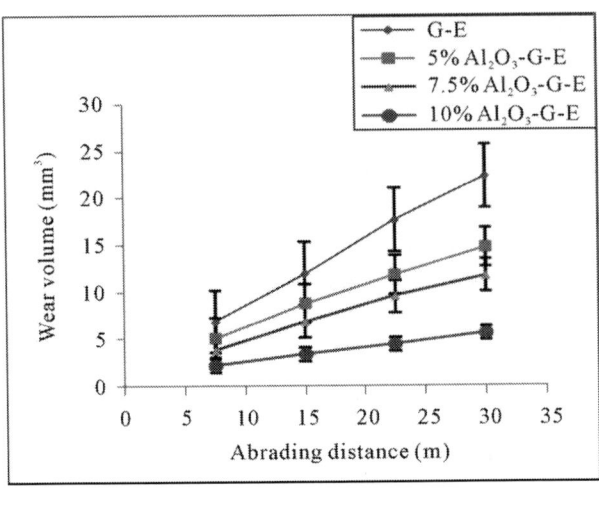

(b)

Figure 3: Variation of wear volume with abrading distance for unfilled and Al_2O_3 filled G-E composites: (a) 320 and (b) 600 grit SiC paper.

As shown in Figures 3(a) and (b), the wear volume of composites is 2.5 - 3.75 times greater than that of unfilled G-E composite. In the specimen worn at a load of 10 N with 320 grit SiC, wear debris did not adhere to the SiC paper. However, in the specimen worn under the same test conditions except the grit size of SiC (600 grit); some abrasive particles penetrated more into the matrix. The fine particles which were detached from the counter surface (SiC paper) fill the cavities and modified the specimen surface. Therefore, the wear volume with 600 grit SiC paper decreased when compared to 320 grit SiC paper. The wear volume loss is less in Al_2O_3 filled G-E composites and it can be attributed to inherent better mechanical properties and spherical shape of Al_2O_3. Also, G-E composites with Al_2O_3 filler addition, improved the mechanical properties listed in the Table 3.

Specific Wear Rate

The variation in the specific wear rate of composites worn on 320 and 600 grit SiC papers at 10 N against abrading distance is shown in

Figures 4(a) and (b) respectively There is a distinct difference between the specific wear rate behaviour.

It is clear from Figures 4(a) and (b) that the specific wear rate for G-E and Al_2O_3 filled G-E composites are increasing with abrading distance and decreased with an increase in the grit size of the SiC paper. This figure also shows that under higher abrading distance (30 m) the specific wear rate for G-E and Al_2O_3 filled G-E composite is following a decreasing trend. Above 15 m abrading distance (severe condition), the specific wear rate for G-E and Al_2O_3 filled G-E composite is following an increasing followed by stable trend. Generally there is the largest drop in specific wear rate for G-E with the addition of Al_2O_3 filler. This behavior can be attributed to the presence of Al_2O_3, which is embedded within the matrix material, covers the packets of plain weave woven glass fabric and impart additional strength to the composite. Generally reinforcements in the form of fibers are sought to increase strength and specific modulus. This is so in conventional static and dynamic tests. In the case of wear, the interaction at the interface between the test specimen and the abrasive paper is a key factor. Lancaster [30] has shown the product of σ and e (where σ is the ultimate tensile strength and e is the ultimate elongation) is a very important factor which controls the abrasive wear behaviour of composites.

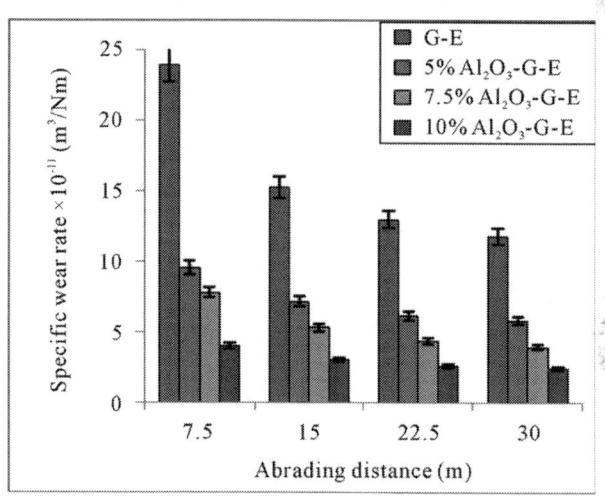

(a)

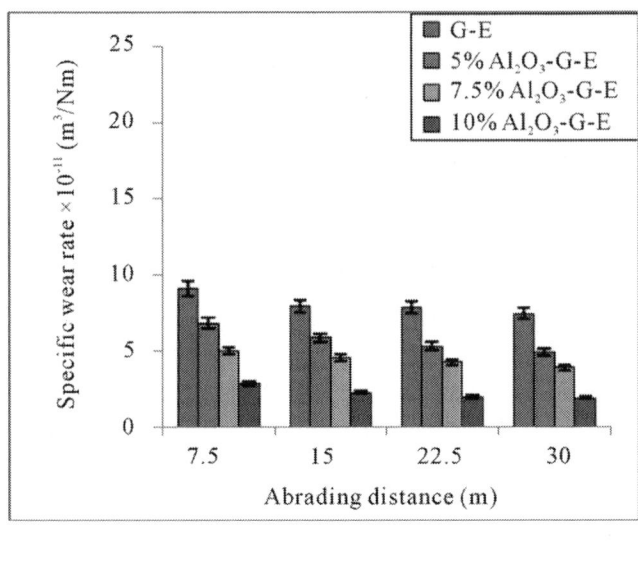

(b)

Figure 4: Variation of specific wear rate with abrading distance for unfilled and Al_2O_3 filled G-E composites: (a) 320 and (b) 600 grit SiC paper.

Generally fiber/filler reinforcement increases the tensile strength (σ) of neat polymer, they usually greatly decrease the ultimate elongation at break (e) and hence the product ($\sigma \times e$) may become smaller than that of neat polymer. Hence, reinforcement usually leads to deterioration in abrasive wear resistance. How these values get changed in the context of filler is a point that needs further investigation. The order of wear resistance behavior of composites is as follows: 10% > 5% > 0% by weight of Al_2O_3.

Worn Surface Morphology (SEM Pictures Analysis)

To correlate the wear data effectively, scanning electron photomicrographs of worn surfaces of G-E and 10 wt% Al_2O_3 filled G-E composite samples are shown in Figures 5(a) and (b) and Figures 6(a) and (b). Several mechanisms have been proposed to explain how material is removed from the surface during abrasion. Because of the complexity of abrasion, no one mechanism completely contributes to all the wear loss. In general, the abrasive wear process involves four different mechanisms namely microploughing, microcutting, microfatigue and microcracking. Using SEM photomicrographs it is possible to identify qualitatively the dominant wear mechanisms under abrasion.

Figure 5(a) shows scanning electron microscope (SEM) micrographs of glass fiber reinforced epoxy samples abraded against 320 grit SiC paper. Figure 5(a) shows some ploughing marks on the surface, matrix damage and exposure of glass fibers. These exposed fibers tend to fracture and their removal from the surface of the composite. The matrix is heavily damaged by ploughing and cutting action by the higher sized SiC particles. Overall surface topography indicated more fiber pulverization, more fiber breakage and less fiber-matrix debonding. The micrograph also indicates the crack propagation of the matrix, deterioration of the fiber-matrix adhesion due to repetitive mechanical stress and some fiber pullout from the matrix is also visible.

Figure 5(b) shows SEM pictures of unfilled G-E samples abraded against 600 grit abrasive papers. Further, few ploughing marks on the surface, matrix damage and very little exposure of glass fibers are seen from the SEM picture. The matrix is damaged more and more microcracks in the matrix are also visible from the micrograph. Further, smooth surface of the matrix and at some regions cracks and also voids are evident from the photomicrograph. This is attributed to the finer abrasive particles get crushed as the abrading distance increases and the SiC particles become ineffective. The SEM picture also indicates the deterioration of the fiber-matrix adhesion due to repetitive mechanical stress and debonding of fibers from the matrix.

Figures 6(a) and (b) show the SEM micrographs of the 10 wt% Al_2O_3 filled G-E composite. Figures 6(a) and (b) exhibit the worn surfaces sliding against SiC paper having a grit size of 320 and 600, respectively. From Figures 6(a) and (b), it can be seen that in the both cases of abrasive grit size of 320 and 600, there exist less wear debris on the worn surfaces, leading to improved wear resistance. On the worn surface of composite abraded against 320 grit SiC paper, the abrasive grooves are narrow and wider Figure 6(a). However, the worn surface of the same sample slide against 600 grit SiC paper indicate shallow and no grooves due to the smaller size of the abrasive particles Figure 6(b).

Comparing the photomicrographs of the composites tested, the extent of damage is less in the case of 10 wt% Al_2O_3 filled G-E composite.

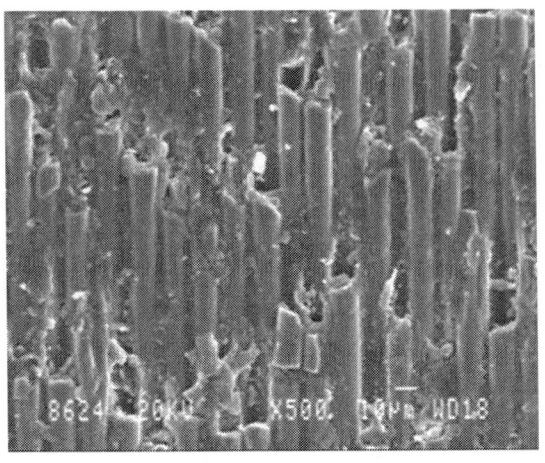

(a)

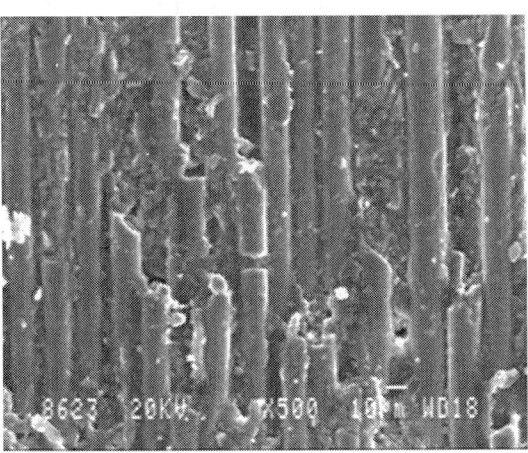

(b)

Figure 5: SEM micrographs of unfilled G-E composite using (a) 320 and (b) 600 grit SiC papers.

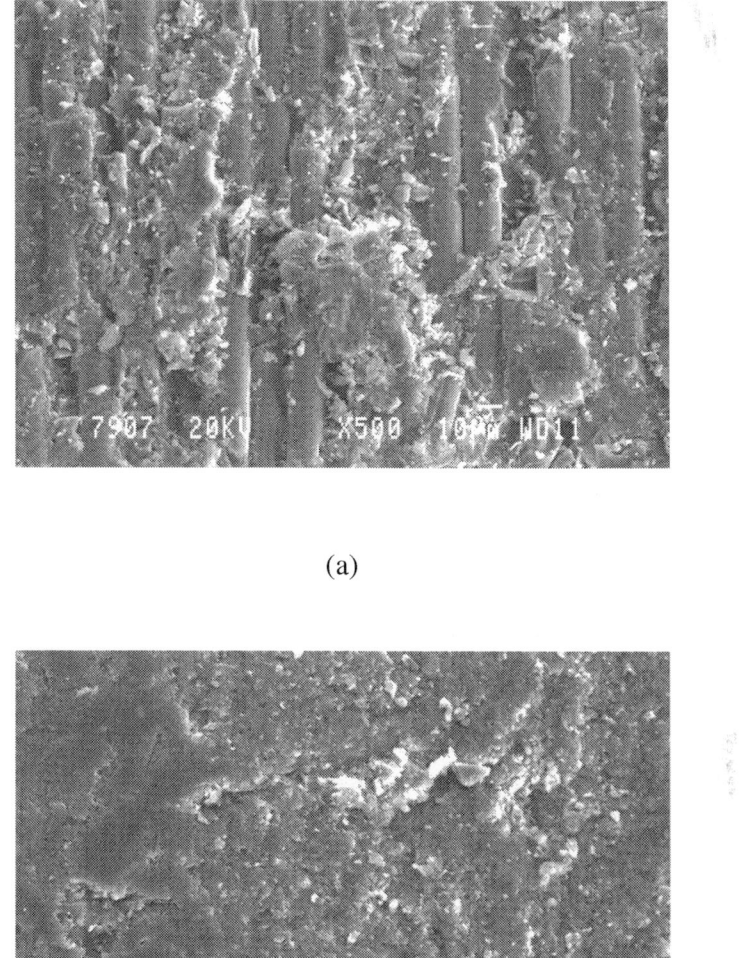

(a)

(b)

Figure 6: SEM micrographs of 10 wt% Al_2O_3 filled G-E composite using (a) 320 and (b) 600 grit SiC papers.

CONCLUSIONS

The mechanical and tribological performances of G-E and Al_2O_3 filled G-E composite were investigated and the following conclusions were drawn.

- The aluminium oxide filler addition to G-E samples has exceptionally improved the abrasive wear performance and the mechanical properties like tensile strength, tensile modulus and hardness properties;
- Two-body wears experimental results showed that the grit size of the abrasive paper greatly affected the wear rate of the composites;
- The wear volume loss increased in glass-epoxy composites with increasing the abrading distance;
- For the specific range of grit size of SiC paper and abrading distance explored in this study, the grit size of SiC paper has shown more influence on the wear behavior of G-E and Al_2O_3 filled G-E composite than the abrading distance;
- In single-pass condition, cutting and multi-pass condition, microcracking and ploughing are the dominant wear mechanisms.

REFERENCES

1. "ASM Handbook," ASM International, Materials Park, Vol. 18, 1992.

2. K. Friedrich, Z. Lu and A. M. Hager, "Recent Advances in Polymer Composites Tribology," Wear, Vol. 190, No. 2, 1996, pp. 139-144 doi: 10.1016/0043-1648(96)80012-3.

3. J. K. Lancaster, "The Effect of Carbon Fiber Reinforcement on Friction and Wear of Polymers," Journal of Physics D: Applied Physics, Vol. 1, No. 5, 1968, pp. 549-555.

4. Z. Lu, K. Friedrich, W. Pannhorst and J. Heinz, "Wear and Friction of Unidirectional Carbon Fiber-Glass Matrix Composite against Various Counterparts," Wear, Vol. 162- 164, 1993, pp. 1103-1110 doi:10.1016/0043-1648(93)90129-A.

5. B. Viswnath, A. P. Verma and C. V. S. Kameswara Rao, "Effect of Reinforcement on Friction and Wear of Fabric Reinforced

Polymer Composites," Wear, Vol. 167, No. 2, 1998, pp. 93-99 doi: 10.1016/0043-1648(93)90313-B.

6. K. H. Zum Gahr, "Wear by Hard Particles," Tribology International, Vol. 31, No. 10, 1998, pp. 587-596 doi: 10.1016/S0301-679X (98)00079-6.

7. I. M. Hutchings, "Mechanism of Wear in Powder Technology: A Review," Powder Technology, Vol. 76, No. 1, 1993, pp. 3-13. doi:10.1016/0032-5910(93)80035-9

8. M. J. Neale and M. Gee, "Guide to Wear Problems and Testing for Industry," William Andrew Publishing, New York, 2001.

9. "Standard Terminology Relating to Wear and Erosion," Annual Book of Standards, ASTM, Vol. 3, 1987, pp. 243- 250.

10. K. N. Shivakumar, G. Swaminathan and M. Sharpe, "Carbon/Vinyl Ester Composites for Enhanced Performance in Marine Applications," Journal of Reinforced Plastics and Composites, Vol. 25, No. 10, 2006, pp. 1101-1116 doi:10.1177/0731684406065194.

11. B. Suresha, G. Chandramohan, P. Sampathkumaran, S. Seetharamu and S. Vynatheya, "Friction and Wear Characteristics of Carbon-Epoxy and Glass-Epoxy Woven Roving Fiber Composites," Journal of Reinforced Plastics and Composites, Vol. 25, No. 7, 2006, pp. 771-782. doi:10.1177/0731684406063540

12. C. Soutis, "Carbon Fiber Reinforced Plastics in Aircraft Construction," Materials Science and Engineering: A, Vol. 412, No. 1-2, 2005, pp. 171-176.doi:10.1016/j.msea.2005.08.064.

13. J. Bijwe, R. Rattan and M. Fahim, "Erosive Wear of Carbon Fabric Reinforced Polyetherimide Composites: Role of Amount of Fabric and Processing Technique," Polymer Composites, Vol. 29, No. 3, 2008, pp. 337-344. doi:10.1002/pc.20544

14. R. Rattan and J. Bijwe, "Carbon Fabric Reinforced Polyetherimide Composites: Influence of Weave of Fabric and Processing Parameters on Performance Properties and Erosive Wear," Materials Science and Engineering: A, Vol. 420, No. 1-2, 2006, pp. 342-350.doi:10.1016/j.msea.2006.01.090.

15. J. L. Abot, A. Yasmin, A. J. Jacobsen and I. M. Daniel, "In-Plane Mechanical, Thermal and Viscoelastic Properties of a Satin Fabric Carbon/Epoxy Composite," Composites Science and

Technology, Vol. 64, No. 2, 2004, pp. 263-268 doi:10.1016/S0266-3538(03)00279-3.

16. R. Park and J. Jang, "The Effects of Hybridization on the Mechanical Performance of Aramid/Polyethylene Intraply Fabric Composites," Composites Science and Technology, Vol. 58, No. 10, 1998, pp. 1621-1628 doi: 10.1016/S0266-3538(97)00228-5.

17. J. Bijwe and R. Rattan, "Influence of Weave of Carbon Fabric in Polyetherimide Composites in Various Wear Situations," Wear, Vol. 263, No. 7-12, 2007, pp. 984-991.doi:10.1016/j.wear.2006.12.030.

18. W. I. J. Zaini, M. Y. A. Fuad, Z. Ismail, M. S. Mansor and J. Mustafah, "The Effect of Filler Content and Size on the Mechanical Properties of Polypropylene/Oil Palm Wood Flour Composite," Polymer International, Vol. 40, No. 1, 1996, pp. 51-55.doi:10.1002/(SICI)1097-0126(199605)40:1<51::AID-PI514>3.0.CO;2-I

19. A. P. Harsha and U. S. Tewari, "Tribo Performance of Polyaryletherketone Composites," Polymer Testing, Vol. 21, No. 6, 2002, pp. 697-702 doi:10.1016/S0142-9418(01)00145-3.

20. B. Suresha, G. Chandramohan, Siddaramaiah, P. Sampathkumaran and S. Seetharamu, "Three-Body Abrasive Wear Behaviour of Carbon and Glass Fiber Reinforced Epoxy Composites," Materials Science and Engineering: A, Vol. 443, No. 1-2, 2007, pp. 285-291. doi:10.1016/j.msea.2006.09.016.

21. D. C. Evans and J. K. Lancaster, "The Wear of Polymers," In: D. Scott, Ed., Treatise on Materials Science and Technology, Academic Press, New York, Vol. 13, 1979, and pp. 85-139.

22. P. H. Shipway and N. K. Ngao, "Microscale Abrasive Wear of Polymeric Materials," Wear, Vol. 255, No. 1-6, 2003, pp. 742-750 doi:10.1016/S0043-1648(03)00106-6.

23. M. Cirino, R. B. Pipes and K. Friedrich, "Evaluation of Polymer Composites for Sliding and Abrasive Wear Applcations," Composites, Vol. 19, No. 5, 1988, pp. 383-392.doi:10.1016/0010-4361(88)90126-7

24. M. Cirino, K. Friedrich and R. B. Pipes, "The Abrasive Wear Behaviour of Continuous Fiber Polymer Composites," Journal of Materials Science, Vol. 22, No. 7, 1987, pp. 235-247. doi:10.1007/BF01082134

25. C. Lhymn, K. E. Tempelmeyer and P. K. Davis, "The Abrasive Wear of Short Fiber Composites," Composites, Vol. 16, No. 2, 1985, pp. 127-136. doi:10.1016/0010-4361(85)90619-6.

26. K. Friedrich "Advances in Composite Technology," In: K. Friedrich and R. B. Pipes, Eds., Elsevier, the Netherlands, Vol. 8, 1993, pp. 209-276.

27. K. Friedrich, "Wear Model for Multiphase Materials and Synergistic Effect in Polymeric Hybrid Composites," In: K. Friedrich and R. B. Pipes, Eds., Advances in Composite Technology, Composite Materials Series, Elsevier, the Netherlands, 1993.

28. J. Bijwe, C. M. Logani and U. S. Tewari, "Influence of Fillers and Fiber Reinforcement on Abrasive Wear Resistance of Some Polymeric Composites," Proceeding of the International Conference on Wear of Materials, Denver, 8-14 April 1989, pp. 75-92.

29. B. Suresha, G. Chandramohan, P. Sampathkumaran and S. Seethuramu, "Investigation of the Friction and Wear Behavior of Glass-Epoxy Composite with and without Graphite Filler," Journal of Reinforced Plastics and Composites, Vol. 26, No. 1, 2007, pp. 81-93.doi:10.1177/0731684407069958.

30. J. K. Lancaster, "Friction and Wear in Polymer Science," In: A. D. Jenkins, Ed., a Material Science Hand Book, North Holland, Amsterdam, 1972.

Numerical Studies on Mechanical Behavior of Adhesive Joints

Xiaocong He, and Yue Zhang

Innovative Manufacturing Research Centre, Kunming University of Science and Technology, Kunming 650500, China

ABSTRACT

This paper describes some finite element models for analyzing the mechanical behavior of adhesive joints. In these models five layers of solid elements were used across the adhesive layer in order to increase the accuracy of the results. The finite elements were refined gradually in steps from adherends to adhesive layer. In these models, most of the adherends and adhesive were modeled using solid brick elements but some solid triangular prism elements were used for a smooth transition. In some of the models, linear interpolation elements of full or reduced integration and of hybrid formulation were used. In other models, quadratic interpolation elements of full or reduced integration and of hybrid formulation were used. Comparisons are drawn between models with different modeling approaches as well as different types of element combinations in order to find a suitable model to predict the behavior of adhesive joints.

INTRODUCTION

Fastening techniques are used extensively in different industry fields for joining various materials in the assembly of components and structures. Many efforts have been spent to develop sheet material joining techniques for application into light-weight structures [1–3].

Adhesive bonding has many characteristics comparable with conventional mechanical fastening and welding methods used in structural engineering. It also has many exclusive advantages such as low bonding temperature, light weight, high stiffness, and good fatigue resistance. Consequently it is becoming a widespread candidate technique for joining light-weight structural components. A considerable amount of theoretical and experimental research has been carried out on the static and dynamic behavior of adhesive joints (e.g., [4, 5]).

To design structural adhesive joints, it is necessary to be able to analyse them. This means determining the stresses and strains under a given loading and predicting the probable points of failure. When different boundary conditions are considered by a closed-form analysis, the limitation is how tractable a realistic mathematical model is within an algebraic solution. Usually it is necessary to simplify the models to some extent to make analytical solutions feasible. Many studies have been published all with analytical or experimental simplifications that restrict the usefulness of the results. With finite element (FE) techniques, however, the limiting factor is more likely to be computing power. The FE method now commonly used is well suited to the estimation of stresses in joints of almost any geometric shape [6–9].

During the last four decades, many of the existing adhesively bonding processes have been simulated by FE methods. Woole and Carver's paper [10] was concerned with the stress analysis by FE method of a bonded single-lap adhesive-bonded joint. A modified version of the well-known Wilson stress analysis program was used in the case of plane stress. The authors used 2 elements' thickness to model the through-the-thickness behavior of the adhesive layer. Stress concentrations as functions of dimensionless, geometric, and material parameters were presented. However, because of the sharp discontinuity between the mechanical properties of the adherend and the adhesive, the use of 2 elements is not sufficient. Smooth transitions between the adherends and the adhesive are necessary in order to obtain accurate results. In later work by Adams and Peppiatt [11], stresses in a standard metal-to-metal adhesive-bonded lap joint were analyzed using a two-dimensional FE method and comparisons were made with previous analyses. In the paper, particular attention was paid to the stresses at the ends of the adhesive layer. Unlike previous work, which assumes the adhesive to have a square edge, the adhesive spew was treated as a triangular fillet. The results show that the highest stresses exist at the adherend corner within the spew. This model is closer to realistic adhesive joints. Carpenter and Barsoum [12] modeled the adherends as two-node beam elements and the adhesive layer as a linear plane element with offset nodes. The number of degrees of freedom is reduced appreciably by this approach because the adherends and the adhesive use the same node.

Anyfantis and Tsouvalis's work [13] was focused on the numerical simulation of single-lap bonded joints, based on cohesive zone modelling techniques. The models were built in a 3D FE space. The adherends were modelled with continuum elements whereas the entire adhesive layer was modelled with cohesive elements. A mixed-mode cohesive model was used as the constitutive relationship between the cohesive elements. The traction increase part of the cohesive laws is given by an exponential function, which describes the elastoplastic adhesive response, and the traction decrease part is given by a linear function, which describes the initiation and propagation of damage. By using this model, it was possible to calculate the developed peel and in-plane and out-of-plane shear stresses over the adhesive area. Hybrid-adhesive joints are an alternative technique for stress reduction in adhesive joints. The joints have two types of adhesives in the overlap region. The stiff adhesive should be located in the middle and the flexible adhesive at the ends. The effect of the hybrid-adhesive bond line on the shear and peeling stresses of a double lap joint was investigated by Özer and Öz [14]. A 3D FE model of the double lap joint has been created based on solid and contact elements. The contact problem was considered by modelling the interface as two surfaces belonging to adherend and adhesive. The results show that the stress components can be optimized using appropriate bond-length ratios.

In the case of analysis of adhesive joints, the thickness of adhesive is much smaller than that of the adherends. FE meshes must accommodate both the small dimension of the adhesive thickness and the larger dimension of the remainder of the whole model. Moreover, the failures of adhesive joints usually occur inside the adhesive layer. In other words, the strength of adhesives is usually lower than that of adherends. It is thus essential to model the adhesive layer by a FE mesh which is smaller than the adhesive thickness. The result is that the FE mesh must be several orders of magnitude more refined in a very small region than is needed in the rest of the joint. It is also important that a smooth transition between the adherends and the adhesive be provided. To determine the physical nature of adhesive joints, many researchers have limited their investigations to single-lap joints because they involve relatively simple and convenient test geometries. However, most other joints may be obtained through some combination or repetition of this basic type.

This paper describes some FE models for analyzing the behavior of single-lap adhesive joints. To overcome the limitations described above, five layers of solid elements were used across the 0.05 mm thick adhesive layer. The main objective of this treatment was to increase the accuracy of the results. The FE models were refined gradually in steps from adherends to adhesive layer. Most of the adherends and the adhesive were modeled using solid brick elements but some solid triangular prism elements were used for a smooth transition. Comparisons are drawn between models with different modeling approaches as well as different types of element combinations in order to find a suitable model to predict the behavior of adhesive-bonded single-lap joints.

CONFIGURATIONS AND MATERIAL PROPERTIES

The single-lap adhesive joint studied in the present paper includes the lower adherend, adhesive layer, and upper adherend, as shown in Figure 1. The two adherends used were 2024-T3 aluminium alloy plates of dimensions 200 mm long × 25 mm wide × 4 mm thickness. The elastic material constants of the adherends were as follows: Poisson's ratio v=0.03 and Young's modulus E=70 GPa. The elastic material constants of the adhesive investigated were Poisson's ratio (V_{ad})=0.30and Young's modulus (E_{ad})=2 GPa.

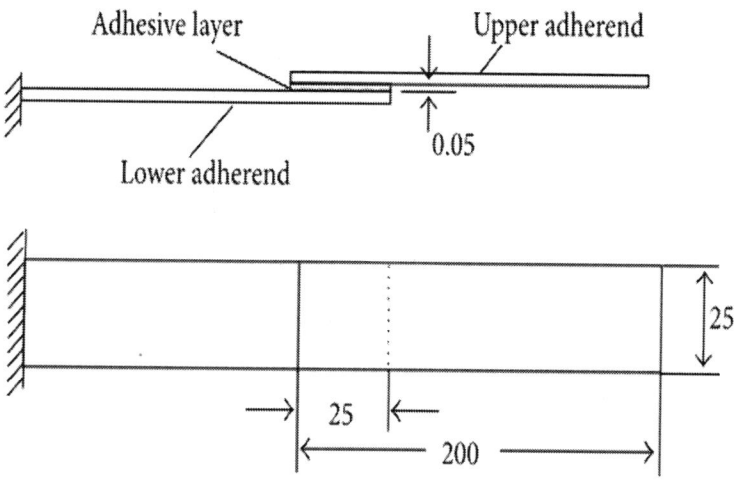

Figure 1: A single-lap adhesively bonded joint (dimensions in mm).

FE MODELS

Model 1 (Coarse Transition Mesh Design)

The FE mesh in model 1 was created using the PATRAN menu-driven FE pre- and postprocessing program operating in an X-window environment. Input into the program was the description of 14920 linear brick and triangular prism elements by indicating the material properties for the elements. The locating of nodal points was accomplished by dividing the configuration into 81 solid models. The original FE mesh of model 1 is shown in Figure 2.

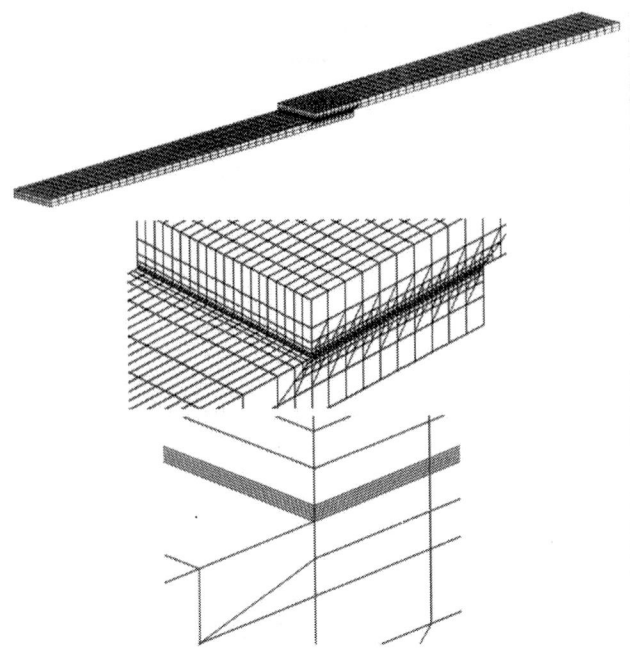

Figure 2: Original FE mesh of model 1.

Most of the geometry of the adherents and the adhesive was modeled using the 8-node solid elements. But at the transition zones from the adherents to the adhesive, where the mesh density is very high, some 6-node transition elements were used. Furthermore, the adhesive layer was divided into 40equal parts along its length (x direction) and 20 equal parts along its width (y direction) in order to obtain an accurate indication of the variation of stresses in the lengthwise (x) and breadthwise (y) directions. Nodal points were located automatically by the PATRAN software as a function of the length and width of the adhesive layer, that is, in accordance with the geometric parameters of the model. Also the material parameters of the adhesive and adherents were input via the PATRAN software.

It can be seen from model 1 that the ratio of the thickness of the adherend elements to the thickness of the adhesive elements is 12.5.This is an abrupt transition in thickness. Also, the ratio of the lengths of the adhesive elements in the x and y directions to their thickness is 62.5 and 125, respectively. The adhesive elements are therefore very long and thin.Thismesh is therefore regarded as a coarse mesh.

Model 2 (Smooth Transition Mesh Design)

The FE mesh in model 2 was created using the ABAQUS FE analysis preprocessing program operating in an X-window environment. It was necessary to define the coordinates of the key nodes and the node number of the key elements in this case. Input into the program was the description of 2700 elements by indicating the material properties for the elements. The original FE mesh of model 2 is shown in Figure 3.

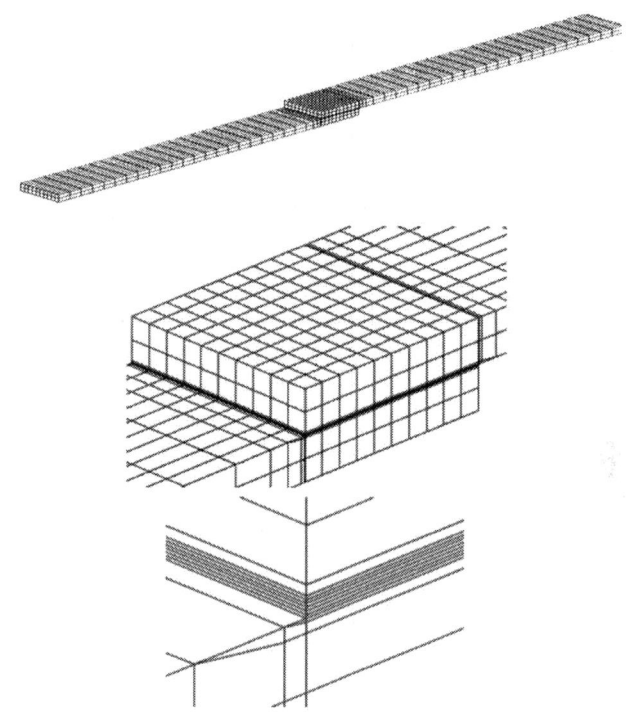

Figure 3: Original FE mesh of model 2.

The geometry of the adherents and adhesive was modeled mainly using the 20-node solid elements. At the transition zones from the adherents to the adhesive, some 15-node transition elements were used. These transition elements were used only in the sections of the adherents which were outside the lap jointed section. The adhesive layer was divided into 10 equal parts along its length (x direction) and

10 equal parts along its width (*y* direction). Nodal points were located by the ABAQUS input file as a function of the length and width of the adhesive layer. The location of these points was in accordance with the geometric parameters of the model. Also the material parameters of the adhesive and adherents were input via the ABAQUS input file.

Model 3 (Smoother Transition Mesh Design)

The FE mesh in model 3 was created also using the ABAQUS FE analysis preprocessing program operating in an X-window environment. Input into the program was the description of 16160 elements by indicating the material properties for the elements. The original FE mesh of model 3 is shown in Figure 4.

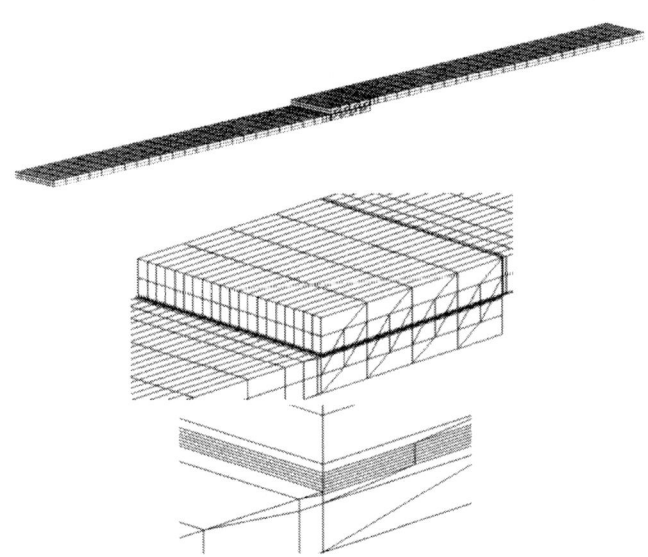

Figure 4: Original FE mesh of model 3.

Most of the adherends and adhesive were also modeled using the 20-node solid elements. But at the transition zones from the adherends to the adhesive, some 15-node transition elements were used. In this case, the transition elements were used in both the lap jointed section of the adherends and the section of the adherends which were outside the lap jointed section. The adhesive layer was divided into 64 equal parts along its length (x direction) and 20 equal parts along its width (y direction) in order to obtain an accurate indication of the variation of stresses in the lengthwise and breadthwise directions. The nodal points were located by the ABAQUS input file as a function of the length and width of the adhesive layer, that is, in accordance with the geometric parameters of the model.

Model 4 (Smoothest Transition Mesh Design)

The FE mesh in model 4 was created using the ABAQUS FE analysis preprocessing program operating in an X-window environment. Input into the program was the description of 57440 elements by indicating the material properties for the elements. The original FE mesh of model 4 is shown in Figure 5.

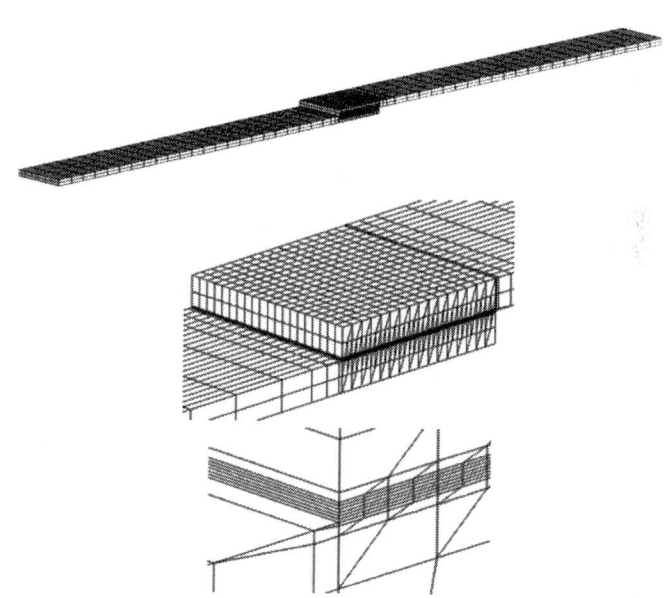

Figure 5: Original FE mesh of model 4.

Again, the adherends and adhesive were mostly modeled using the 20-node solid elements. However, at the transition zones from the adherends to the adhesive, some 15-node transition elements were used. The adhesive layer was divided into 256 equal parts along its length (x direction) and 20 equal parts along its width (y direction) in order to obtain an accurate indication of the variation of stresses in the direction normal to the bond line. As in the previous cases, the nodal points and the material parameters of the adhesive and adherends were input via the ABAQUS input file.

Comparison of FE Models

In order to illustrate the influence of the choice of FE model on the prediction of the mechanical behavior of adhesive joints, comparisons were performed between models with different modeling approaches. Since the failures of adhesive joints usually occur inside the adhesive layer, then only the lap jointed section is of interest.

It is easy to create a FE mesh using the PATRAN software as the nodal points are located automatically. The number of nodal points was reduced by this approach because part of the surface of the adherends and the surface of the adhesive uses the same node. In model 1, for example, input into the program was the location of 26922 nodal points but output of the analysis result was only 16968 nodal points. As a result, unfortunately, the number of nodal points was arranged discontinuously by the program. This makes postprocessing of the FE analysis results difficult. In fact it was found that it is better to use the ABAQUS FE preprocessing program to create the FE mesh of the single-lap adhesive joint.

Unlike model 1, model 2 was created using the ABAQUS FE analysis preprocessing program. This model has a limited number of elements and nodes. In the lap joint section, smooth transitions between the adherends and the adhesive in the z direction were provided. Obviously the adhesive layer needs to be divided into more equal parts in order to obtain an accurate indication of the variation of stresses in the x and y directions.

It can be seen from Figure 5 that, in model 4, the FE model was refined gradually in steps from the adherends to the adhesive layer. That means smooth transitions were provided between the adherends and the adhesive in both the x and z directions. In the y direction, the model was divided into more equal parts than in model 2. Therefore, model 4 was expected to provide more accurate analysis results. However, the disadvantage is that model 4 needs more computing time as it has a larger number of elements and nodes.

In the case of model 3, smooth transitions between the adherends and the adhesive were also provided in both x and z directions. In addition, model 3 has a moderate number of elements and nodes. Thus, model 3 was expected to be the most cost-effective of the 4 models studied.

In order to confirm this point, the stress distributions of the 4 models under tension were investigated. A distributed load of 1000 N was applied at the right end face of the upper adherend in the x direction. This distributed load does not refer to any load condition in particular and is used simply as an example for comparisons between different modeling approaches and different combinations of elements. The boundary conditions of the joint are shown in Figure 6. Because five layers of solid elements were used across the adhesive thickness, a total of six interfaces were obtained. The lower interface, which is between the lower adherend and the adhesive, is denoted by interface 1. Similarly, the upper interface, which is between the adhesive and upper adherend, is denoted by interface 6. The intermediate interfaces are denoted by interfaces 2 to 5.

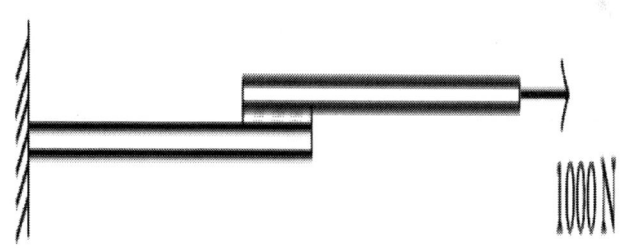

Figure 6: Boundary condition of a single-lap adhesively bonded joint.

Figures 7 and 8 show the distributions of normal stress S11 predicted by the 4 models at interface 1 of the adhesive layer. The dimensions in the x and y directions are displayed in no dimensional form as x/c and y/b where b is the width of joint and c is the length of the bonded section. Figure 7 shows the stress distributions at the front edge ($y/b =$

0) of interface 1. From the symmetry of the y direction, it is clear that the stress distributions at the rear edge ($y/b = 1$) of interface 1 are the same as that at the front edge. Figure 8 shows the stress distributions at the center line ($y/b = 0.5$) of interface 1.

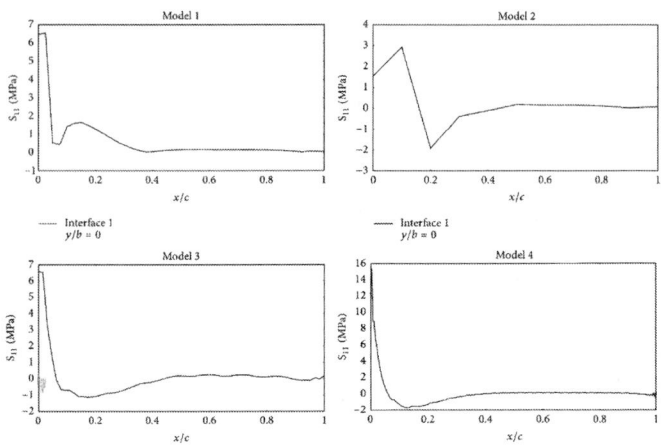

Figure 7: Distributions of S_{11} at the front edge of interface 1 in different models.

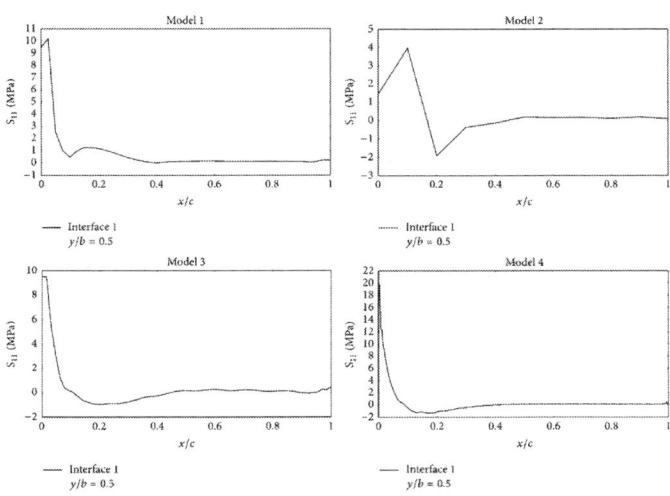

Figure 8: Distributions of S_{11} at the center line of interface 1 in different models.

From Figures 7 and 8, it can be seen that in the x direction the left hand region is subjected to much higher stresses than the right hand region. In the case of model 1, the stress distribution curve does not extend as expected, possibly because the transition mesh is coarse. In the case of model 2, there is a severe oscillation in the stress distribution curves. The stress distribution curves of models 3 and 4, however, extend smoothly. Comparing the predicted direct stress for these models, it can be seen that the stress distributions of models 3 and 4 are similar. Therefore it can be inferred that the adhesive joint is better represented by models 3 and 4. The anomalies observed in stress distributions obtained using models 1 and 2 show that these two models are not adequate for modeling the adhesive joint. The figures show that the predicted results improve in accuracy as the model size increases. The worst result is obtained using model 1 which has a coarse FE mesh, whereas the best result is obtained using model 4 which has the finest FE mesh. Of course model 4 is more accurate in predicting the results than model 3, but it needs much longer CPU time. We conclude, therefore, that model 3 is the most cost-effective choice. Moreover, the results support a preference for choosing finer elements, for example, 20-node elements, to improve the accuracy of prediction rather than choosing more complex models which need much longer CPU time.

ELEMENT TYPES AND TOPOLOGY

Stress/displacement elements were chosen for this study because they are suitable for modeling linear or complex nonlinear mechanical analyses that may involve contact, plasticity, and large deformations [15]. There are different types of stress/displacement elements, which are appropriate for different types of analysis. In the case of analysis of adhesive joints, the elements must accommodate the material properties and joint dimensions of both the adherend and the adhesive. The element combinations are defined in pairs to denote the element types used to model the adherend and adhesive, respectively. The first-order element combinations investigated in the present study include C3D8-C3D8, C3D8-C3D8H, C3D8R-C3D8H, C3D8R-C3D8R, and C3D8R-C3D8RH element combinations. The second-order element combinations investigated in the present study include C3D20-

C3D20, C3D20-C3D20H, C3D20R-C3D20H, C3D20R-C3D20R, and C3D20R-C3D20RH element combinations.

DISCUSSION OF RESULTS

A recent study by the present author [16] showed that the spatial distributions of all the 6 components of stress are similar for different interfaces even though the stress values are slightly different. Since the failure of single-lap bonded joints initiates where high stresses occur, we are only interested in the maximum stresses. The overwhelming majority of maximum stresses occur at interface 1 though a few occur between interfaces 1 and 2 and at interface 6. Furthermore, the maximum stresses at interface 1 are much bigger. This section describes the predicted stress distributions obtained using some 3D stress/ displacement element combinations to model a single-lap adhesive joint under tension.

Stress Distributions Using First-Order Element Combinations

As stated previously, the first-order element combinations investigated in the present study include C3D8-C3D8, C3D8-C3D8H, C3D8R-C3D8H, C3D8R-C3D8R, and C3D8R-C3D8RH element combinations. The stress distributions corresponding to different element combinations were obtained. However, only a few typical distributions will be discussed here. Figure 9 shows the distributions of the 6 components of stress for the C3D8R-C3D8R element combinations at interface 1 as an illustration of the typical 3D stress distribution in the adhesive layer of the first-order element combinations. To enable easy comparison of these stress distributions, all the 6 components of stress are drawn using the same coordinate scales.

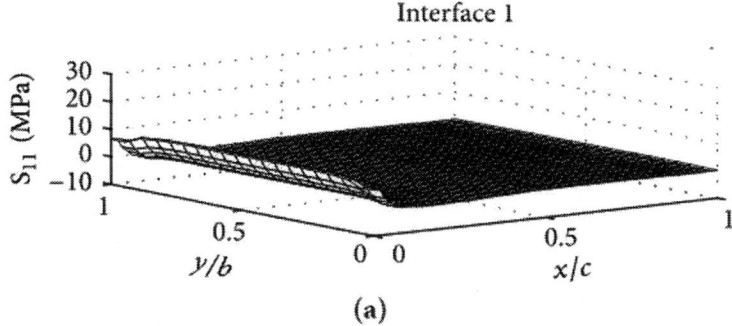

(a)

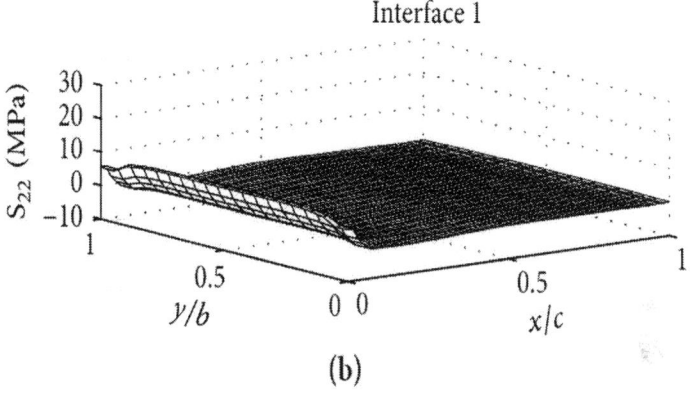

(b)

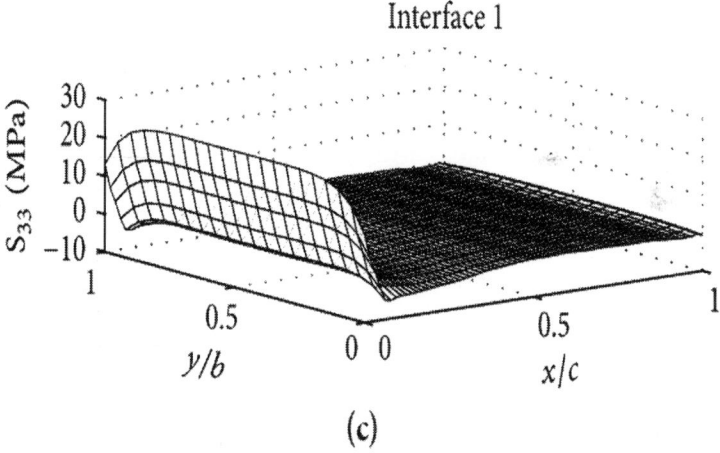

(c)

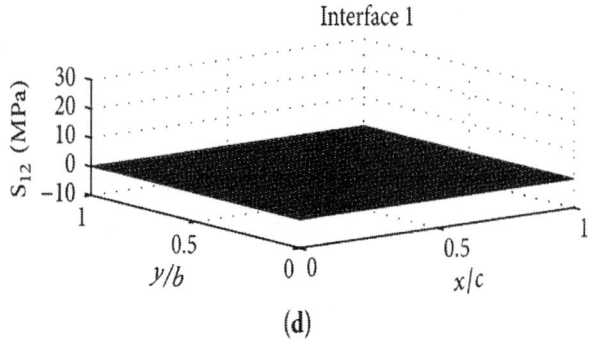

(d)

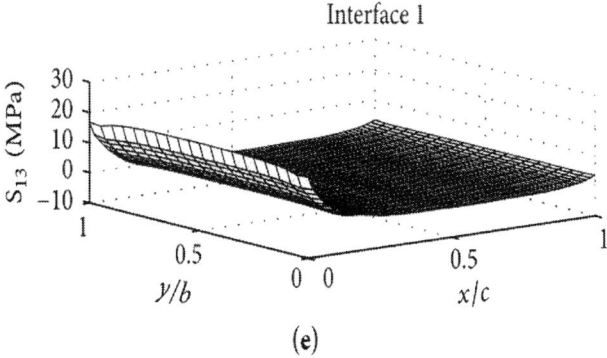

(e)

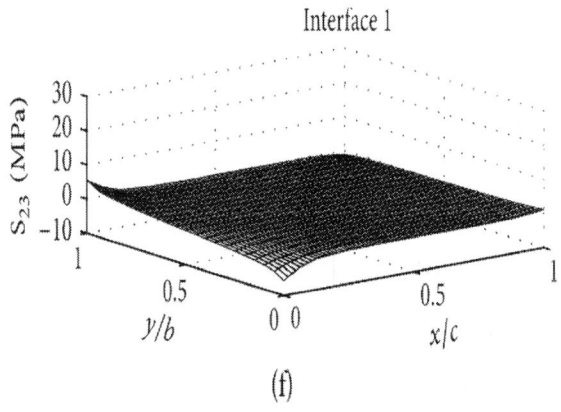

(f)

Figure 9: Distributions of the 6 components of stress in C3D8R-C3D8R element combinations.

The figure shows that the highest stresses are concentrated near the left edge ($x/c = 0$) of the adhesive layer. The S_{33} has the highest magnitude of stress while the S_{12} has the least magnitude. The stress distributions of other first-order element combinations were omitted because they look similar to Figure 9, though there are some distinctions between them. These distinctions are discussed by using two-dimensional plots.

Figures 10 and 11 show two-dimensional plots of the maximum values for the 6 stress components of C3D8-C3D8, C3D8R-C3D8H, and C3D8R-C3D8R element combinations against the no dimensional distances x/c and y/b, respectively. In order to make the figures clear, the following codes are used to denote the element combinations and interfaces in Figures 10 and 11:

8-8: C3D8-C3D8 element combinations,

8R-8H: C3D8R-C3D8H element combinations,

8R-8R: C3D8R-C3D8R element combinations,

Int1: interface 1,

Int5: interface 5,

Int6: interface 6.

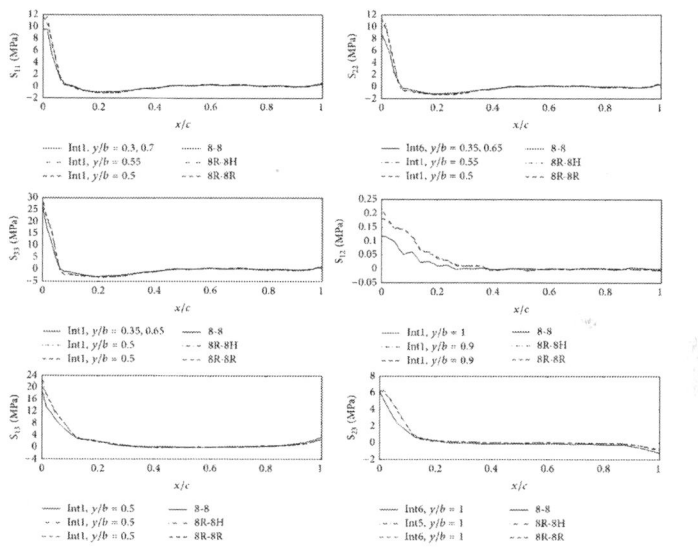

Figure 10: Maximum stresses of the first-order element combinations versus x/c.

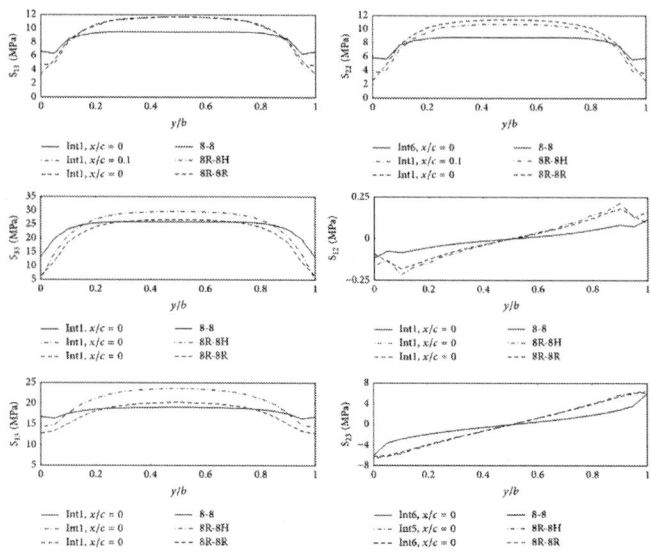

Figure 11: Maximum stresses of the first-order element combinations versus y/b.

The C3D8 element is an 8-node linear brick element. In the case of the C3D8-C3D8 element combinations, most of the maximum stresses (S_{11}, S_{33}, S_{12}, and S_{13}) occur at interface 1. The maximum stresses of S_{22} and S_{23}, however, occur at interface 6. In addition, all 6 maximum Similarly, the C3D8R element is an 8-node linear brick, reduced integration with an hourglass control element, while the C3D8H element is an 8-node linear brick, hybrid with a constant pressure element. In the case of the C3D8R-C3D8H element combinations, S_{11} max, S_{22} max, S_{33} max, S_{12} max, and S_{13} max occur near the center of the left edge of interface 1, while S_{23} max occurs at interface 5.

In the case of the C3D8R-C3D8R element combinations, most of the maximum values of the 6 components of stress occur at the center of the left end of interface 1 except S_{12} max which occurs near the left-rear corner of the interface 1 and S_{23} max which occurs at the left-rear corner of the interface 6.

From Figure 10, it is obvious that the distributions of the direct or normal stresses S_{11}, S_{22}, and S_{33} are similar. The magnitudes of S_{11} and S_{22} are almost identical but the magnitude of S_{33} is more than double the magnitudes of S_{11} and S_{22}. While the distributions of the shear stresses S_{12}, S_{13}, and S_{23} are fairly similar, their magnitudes are widely different. Furthermore, the figures show that the stress distributions predicted by the 8R-8H and the 8R-8R element combinations are very closely correlated. But the stress distribution predicted by the 8-8 element combinations deviates significantly from predictions for the other element combinations. Similarly, it can be seen from Figure 11 that the stress distributions predicted by the 8-8 element combinations are quite different from those predicted by the 8R-8H and the 8R-8R element combinations, the predictions for which are fairly similar. We conclude, therefore, that the 8-8 element combination is not suitable for the analysis of single-lap adhesive joints.

Stress Distributions Using Second-Order Element Combinations

The second-order element combinations investigated in the present study include C3D20-C3D20, C3D20-C3D20H, C3D20R-C3D20H, C3D20R-C3D20R, and C3D20R-C3D20RH element combinations. Figure 12shows the distributions of the 6 components of stress for the C3D20R-C3D20R element combinations at interface 1 as an illustration of the typical 3D stress distribution in the adhesive layer of the second-order element combinations. It can be seen that the highest stresses are concentrated near the left edge ($x/c = 0$) of the adhesive layer. Also, S_{33} has the highest magnitude of stress whereas S_{12} has the least. These observations are identical with those made previously from Figure 9.

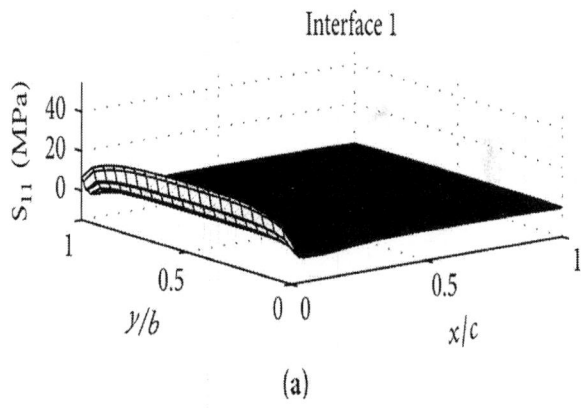

(a)

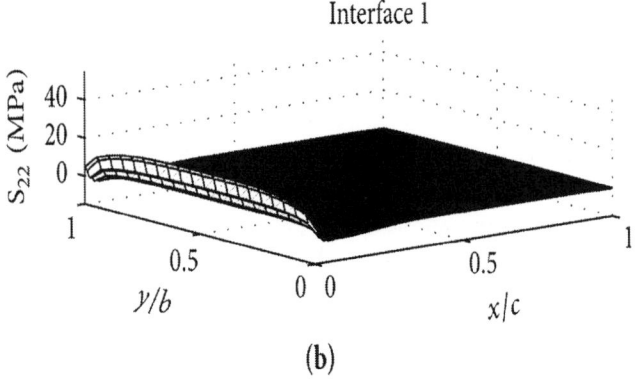

(b)

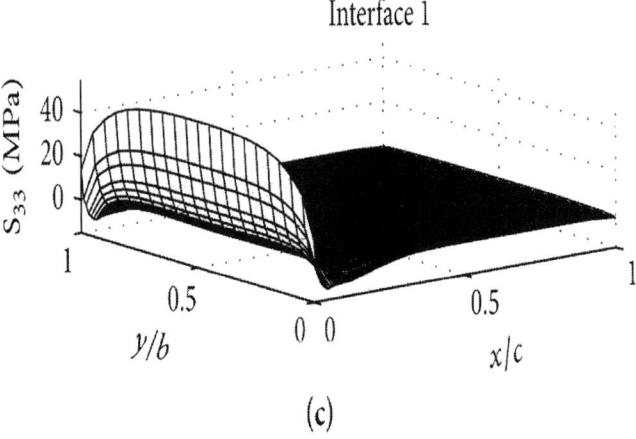

(c)

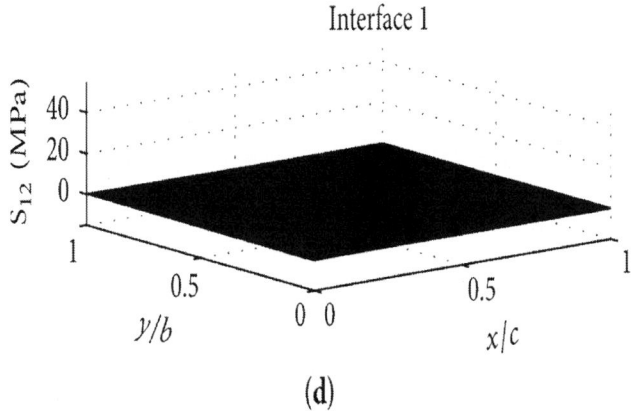

(d)

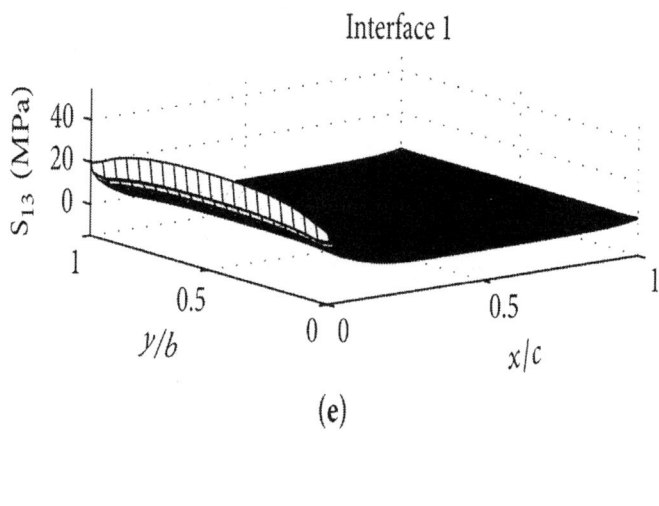

(e)

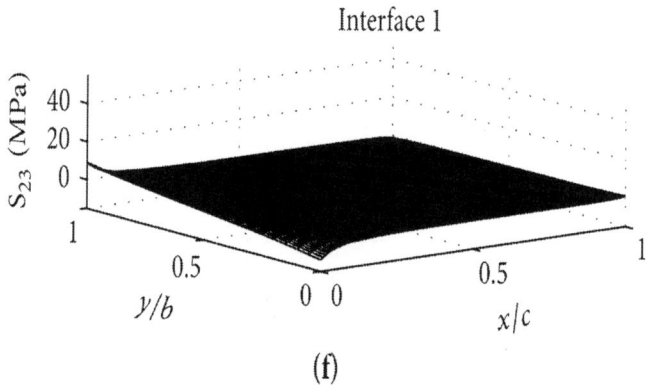

(f)

Figure 12: Distributions of the 6 components of stress in C3D20R-C3D20R element combinations.

Figures 13 and 14 show two-dimensional plots of the maximum values of the 6 stresses components of the C3D20-C3D20, C3D20R-C3D20H, and C3D20R-C3D20R element combinations against the no dimensional distances x/c and y/b, respectively. The stress distributions for the C3D20-C3D20H and C3D20R-C3D20HR element combinations are not included in these figures because they are very close to those of the C3D20-C3D20 and C3D20R-C3D20H element combinations, respectively. The following codes are used to denote the element combinations and interfaces in Figures 13 and 14:

20-20: C3D20-C3D20 element combinations,

20R-20H: C3D20R-C3D20H element combinations,

20R-20R: C3D20R-C3D20R element combinations,

Int1: interface 1,

Int6: interface 6.

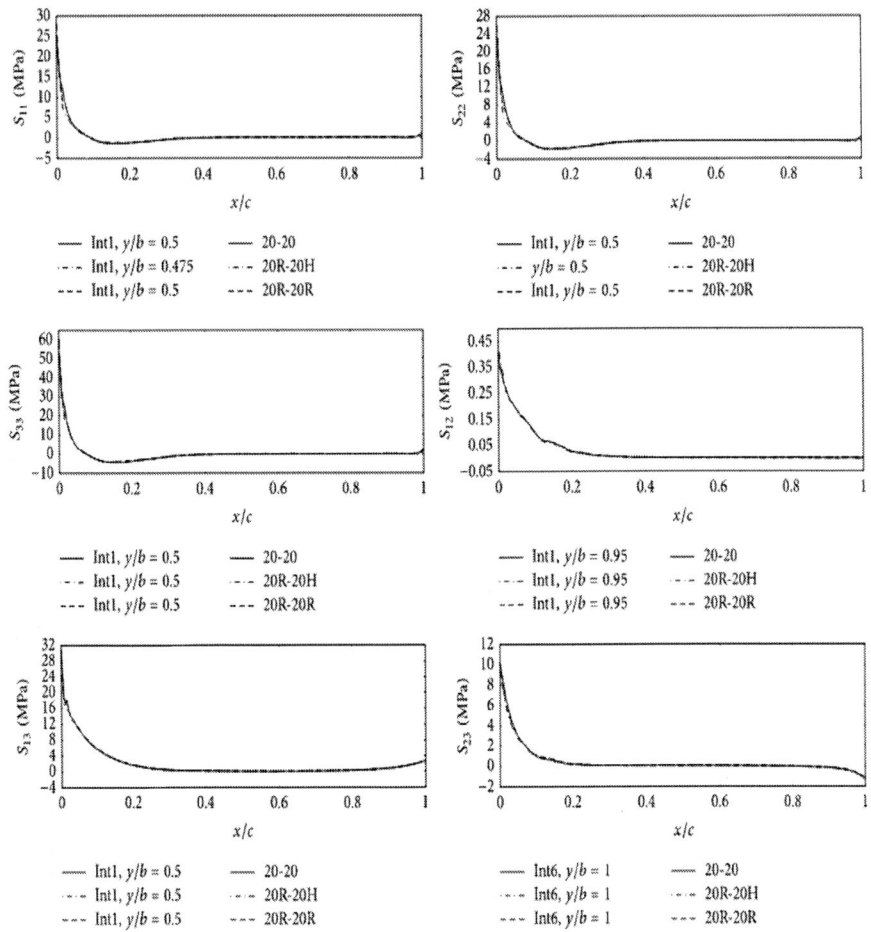

Figure 13: Maximum stresses of the second-order element combinations versus x/c.

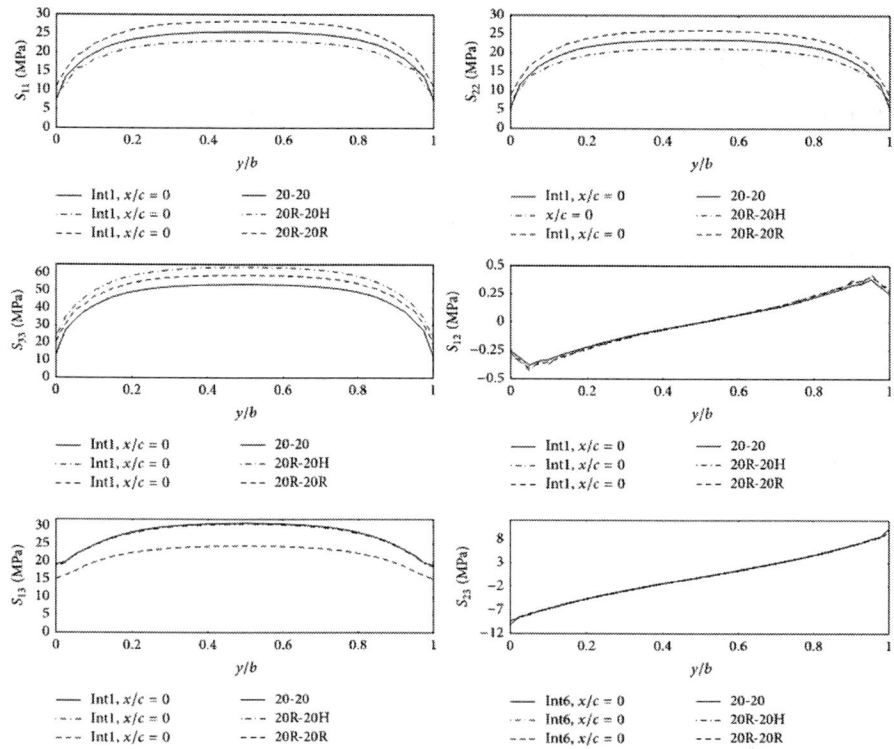

Figure 14: Maximum stresses of the second-order element combinations versus y/b.

The C3D20 element is a 20-node linear brick element. In the case of the C3D20-C3D20 element combinations, it can be seen that most maximum stresses occur at the interface 1 except the $S_{23\ max}$ which occurs at the interface 6.

The C3D20R element is a 20-node quadratic brick, reduced integration element, while the C3D20H element is a 20-node quadratic brick hybrid with a linear pressure element. In the case of the C3D20R-C3D20H element combinations, the $S_{11\ max}$ occurs near the center of the left region. The $S_{22\ max}$ does not occur at an interface but between interfaces 1 and 2. In addition, there is a severe oscillation in the S_{13} curve.

In the case of the C3D20R-C3D20R element combinations, most of the maximum values of the 6 components of stresses occur in the center of the left end of interface 1. The shear stress $S_{12\,max}$ ($x/c = 0$, y/b = 0.95) occurs at the leftrear corner of interface 1 while $S_{23\,max}$ ($x/c = 0$, $y/b = 1$) occurs at the left-rear corner of interface6. It is alsoclear from figures that the stress distribution curves extend smoothly.

Comparison of Maximum Stresses Predicated by Linear and Quadratic Elements

In this section, comparisons are performed between the 8-node element groups and the 20-node element groups. Table 1 shows the maximum values of the 6 components of stress in the 8-node element combinations. It is clear that the stress state in this case is mainly dominated by the normal stress component S_{33} and then the shear stress component S_{13}. Surprisingly, the maximum values of the stress components S_{13} and S_{33} are higher than the maximum value of the stress component S_{11}. The latter would have been expected to be the most dominant since the joint is subjected to tensile loading. The departure from expected behavior is due to the effect of bending at the bonded section of the lap joint. In the 8-node element groups, not only the C3D8-C3D8 and C3D8-C3D8H element combinations, but also the C3D8R-C3D8R and C3D8R-C3D8RH element combinations have exactly the same stress distribution. This observation seems to suggest that the 8-node hybrid elements do not work in the analysis of the single-lap adhesive joints. In addition, the magnitudes of stresses of the 8-node element combinations oscillate in values along the lengthwise direction.

Table 1: Maximum values of stress of 8-node elements combinations

Element combinations	Maximum values of stress (MPa)					
	S_{11}	S_{22}	S_{33}	S_{12}	S_{13}	S_{23}
C3D8-C3D8	9.53	8.82	25.90	0.12	19.07	6.10
C3D8-C3D8H	9.53	8.82	25.90	0.12	19.07	6.10
C3D8R-C3D8H	11.68	10.75	29.69	0.22	23.60	6.63
C3D8R-C3D8R	11.69	11.39	26.72	0.18	20.26	6.28
C3D8R-C3D8RH	11.69	11.39	26.72	0.18	20.26	6.28

Table 2 shows the maximum values of the 6 components of stress of the 20-node element combinations. The stress components S_{13} and S_{33} mainly dominate the stress state as for the 8-node element groups, but the stress values are larger than those of the 8-node element groups. Unlike the 8-node element groups, however, every combination has a different stress distribution. For example, Table 1 shows that the maximum values of the 6 stress components predicated by the C3D8-C3D8 and C3D8R-C3D8R element combinations are identical to those predicated by the C3D8-C3D8H and C3D8R-C3D8RH element combinations, respectively. However, Table 2 shows that the maximum values of the 6 stress components predicated by the second-order element combinations are all different. Also, for the C3D20-C3D20H, C3D20R-C3D20H, and C3D20R-C3D20RH element combinations, the maximum values of normal stress component S_{22} do not occur at an interface but between interfaces 1 and 2. This observation seems to suggest that the 20-node hybrid elements do not work well in this study. This is not very surprising because the adhesive used is very stiff and therefore has a Poisson ratio less than 0.5. Thus the adhesive is not incompressible and the hybrid elements formulated for incompressible or nearly incompressible materials will not work well.

Table 2: Maximum values of stress of 20-node elements combinations

Element combinations	Maximum values of stress (MPa)					
	S_{11}	S_{22}	S_{33}	S_{12}	S_{13}	S_{23}
C3D20-C3D20	25.43	23.49	53.38	0.38	30.85	10.16
C3D20-C3D20H	20.19	18.41	51.85	0.38	30.72	10.20
C3D20R-C3D20H	23.03	21.15	63.51	0.44	30.58	9.99
C3D20R-C3D20R	28.19	25.95	58.82	0.41	24.46	9.43
C3D20R-C3D20RH	25.06	23.81	57.44	0.41	24.45	9.42

Also, it was shown previously that the C3D8-C3D8 element combination is not suitable. Thus, the reasonable choice should be between the following three types of element combinations: C3D8R-C3D8R, C3D20-C3D20, and C3D20R-C3D20R.

As mentioned before, second-order elements provide higher accuracy than first-order elements. They capture stress concentrations more effectively and are better for modeling geometric features.

In addition, second-order elements are very effective in bending-dominated problems.

Reduced integration uses a lower order integration to form the element stiffness and so reduces the run time, especially in three dimensions. For example, element type C3D20R has 8 integration points while C3D20 has 27. Therefore, element assembly is roughly 3.5 times less costly for C3D20R than for C3D20. In addition, second-order reduced integration elements generally yield more accurate results than the corresponding fully integrated elements.

Finally, C3D20R-C3D20R element combinations would be the best element combinations for the analysis of single-lap adhesive joints.

CONCLUSIONS

Some FE models for analyzing the behavior of adhesive joints were described in this paper. In these models five layers of solid elements were used across the adhesive layer which was only 0.05 mm thick, in order to obtain accurate results. The FE models were refined gradually in steps from adherends to adhesive layer. Most of the adherends and the adhesive were modeled using quadratic solid elements but some triangular solid elements were used to give a smooth transition. Comparisons were performed between models with different modeling approaches as well as different types of element combinations.

From the comparisons between the FE models, it is clear that of the 4 models presented in this study model 3 is the most cost-effective. This is because it has a moderate number of elements and nodes and a smooth transition between the adherends and the adhesive in both x and z directions.

The results of the analysis also show that the linear, fully integrated, and hybrid elements are not suitable for the analysis of single-lap adhesive joints. In addition, second-order reduced integration elements provide higher efficiency and accuracy than the corresponding first-order elements.

ACKNOWLEDGMENTS

This study is partially supported by National Natural Science Foundation of China (Grant no. 50965009) and Special Program of the Ministry of Science and Technology, China (Grant no. 2012ZX04012-031).

REFERENCES

1. X. He, F. Gu, and A. Ball, "A review of numerical analysis of friction stir welding," Progress in Materials Science, vol. 65, pp. 1–66, 2014. · ·

2. J. Mucha, "The analysis of lock forming mechanism in the clinching joint," Materials & Design, vol. 32, no. 10, pp. 4943–4954, 2011. · ·

3. X. He, "Finite element analysis of laser welding: a state of art review," Materials and Manufacturing Processes, vol. 27, no. 12, pp. 1354–1365, 2012. · ·

4. X. He and Y. Wang, "An analytical model for predicting the stress distributions within single-lap adhesively bonded beams," Advances in Materials Science and Engineering, vol. 2014, Article ID 346379, 5 pages, 2014. ·

5. X. He, "Bond thickness effects upon dynamic behaviour in adhesive joints," Advanced Materials Research, vol. 97–101, pp. 3920–3923, 2010. · ·

6. I. T. Pearson and J. T. Mottram, "A finite element modelling methodology for the non-linear stiffness evaluation of adhesively bonded single lap-joints: part 1. Evaluation of key parameters," Computers & Structures, vol. 90-91, no. 1, pp. 89–96, 2012. · ·

7. I. T. Pearson and J. Toby Mottram, "A finite element modelling methodology for the non-linear stiffness evaluation of adhesively bonded single lap-joints: Part 2. Novel shell mesh to minimise analysis time,"Computers and Structures, vol. 90-91, no. 1, pp. 76–88, 2012. · ·

8. X. He, "Numerical and experimental investigations of the dynamic response of bonded beams with a single-lap joint," International Journal of Adhesion and Adhesives, vol. 37, pp. 79–85, 2012. · ·

9. X. He, "Dynamic behaviour of single lap-jointed cantilevered beams," Key Engineering Materials, vol. 413-414, pp. 733–740, 2009. · ·

10. G. R. Woole and D. R. Carver, "Stress concentration factors for bonded lap joints," Journal of Aircraft, vol. 8, no. 10, pp. 817–820, 1971. · ·

11. R. D. Adams and N. A. Peppiatt, "Stress analysis of adhesive-bonded lap joints," The Journal of Strain Analysis for Engineering Design, vol. 9, no. 3, pp. 185–196, 1974.

12. W. C. Carpenter and R. Barsoum, "Two finite elements for modeling the adhesive in bonded configurations," Journal of Adhesion, vol. 30, pp. 25–46, 1989. ·

13. K. N. Anyfantis and N. G. Tsouvalis, "Loading and fracture response of CFRP-to-steel adhesively bonded joints with thick adherents—part II: numerical simulation," Composite Structures, vol. 96, pp. 858–868, 2013. · ·

14. H. Özer and Ö. Öz, "Three dimensional finite element analysis of bi-adhesively bonded double lap joint," International Journal of Adhesion and Adhesives, vol. 37, pp. 50–55, 2012. · ·

15. ABAQUS Theory Manual, HKS, Leicester, UK, 1998.

16. X. He, "Influence of boundary conditions on stress distributions in a single-lap adhesively bonded joint," International Journal of Adhesion and Adhesives, vol. 53, pp. 34–43, 2014. ·

Chapter 7

Review: Development of Performance-Based Fire Design for Cold-Formed Steel

Jean C Batista Abreu[1], Luiz M C Vieira[2], Metwally H Abu-Hamd[3], and Benjamin W Schafer[4]

[1]Department of Civil Engineering, Johns Hopkins University, Baltimore, MD, USA

[2]Department of Structural Engineering, University of Campinas, Campinas, Brazil

[3]Structural Engineering Department, Cairo University University, Cairo, Egypt

[4]Department of Civil Engineering, Johns Hopkins University, Baltimore, MD, USA

ABSTRACT

Performance-based fire design for cold-formed steel systems is in its infancy. This paper brings together existing research on cold-formed steel materials, members, and assemblages at elevated temperatures;

and complementary analysis and design methods necessary for the development of analysis-based design for cold-formed steel systems under fire. Cold-formed steel systems have become popular in building construction as both load-bearing and non-load-bearing elements, primarily due to their high strength-to-weight ratio and ease of construction. Consequently, design specifications, and structural analysis tools have rapidly evolved to facilitate engineering design of these complex thin-walled members. However, in fires the performance of cold-formed steel systems are assured by prescriptive detailing and standardized testing. Today, engineering knowledge is rapidly advancing, providing the opportunity to contemplate analysis-based design as an enabling tool for general performance-based fire engineering of cold-formed steel systems. The review provided here includes experimental results on mechanical and thermal properties of cold-formed steel and temperature dependent constitutive relations, subsystem testing and computational simulations, and analysis models and exploratory methods for fire design, i.e., the building blocks towards performance-based fire design for cold-formed steel systems.

INTRODUCTION

Cold-formed steel (CFS) members are manufactured from cold bent sheet steel, approximately from 0.5 mm to 3.0 mm thick. The most common members are channels (tracks) and lipped channels (studs and joists). CFS stud and track are used extensively in buildings as the framing for interior partition walls, exterior curtain walls, and more recently as the complete load-bearing system (Allen[2004], Schafer [2011]). CFS interior partition walls are framed with studs, have track at top and bottom, and are then sheathed (most commonly) with gypsum wallboard(s). In a fire, partition walls serve as primary barriers to maintain building integrity, and avoid the spread of fire between compartments (rooms). In the United States, the assemblies are required to be ¿fire-rated¿ (IBC [2012]) based on their ability to withstand a standardized ¿fire¿ test (ASTM [2012]). The fire-resistance rating is expressed by the number of hours that the assembly can maintain its integrity while containing gases and excessive temperature increases out of the fire compartment. A large number of assemblages have been tested, and industry has assembled catalogs of the prescriptive details that can provide a given fire rating (CFSEI [2012]). These prescriptive solutions are critical to current design and represent an important review of the state of the art in their own right; however, the focus here is on enabling performance-based fire design, not additional prescriptive solutions.

To date, fire design for load-bearing CFS systems (where the complete structural system is framed from CFS members) has followed the same test-based, prescriptive detail-driven approach that has been previously established for interior partition walls. However, given the wide variety of possible members and details the prescriptive approach has several drawbacks, as discussed in the case for performance-based fire design of cold-formed steel systems section. In addition, hot-rolled steel has demonstrated the possibilities and advantages of enabling performance-based fire design (e.g. AISC[2010]). Further, recent research has taken the first steps towards performance-based design for establishing the fire resistance of CFS structures, including temperature dependence of the material (mechanical and thermal) behavior, thermo-mechanical response of members and sub-systems, and temperature dependence of member strength predictions. This state-of-the-art review discusses current research and recent findings on the fire performance of CFS.

Fire Demands and Heat Transfer Analysis of CFS Systems

Fundamental to determination of the fire resistance is establishing the fire demand and then propagating that demand to the underlying members. Ideally, performance-based fire design brings the demand (fire modeling), propagation (heat transfer), and capacity (strength at elevated temperatures) all into the realm of analysis. In such a situation, the complete system may be designed for the desired fire performance with interactions between demand, propagation, and capacity fully included through analysis. Although the focus of this review is on capacity, demand and heat transfer is briefly reviewed here to establish the conditions under which the capacity is evaluated.

Fire Demand

One of the first formal attempts to account for fire action on building structures emerged in 1918, when the ASTM standardized a time-temperature relationship (called the fire curve) to consistently evaluate the fire resistance of buildings. The fire curve was intended to represent a worst-case expected fire scenario, based on empirical data from timber construction (ASCE [2009]). Similar time-temperature relationships have been implemented internationally. Typically, the fire curve is only weakly related to the actual time-temperature curve for a fire in a modern building. However, standard fire curves provide a consistent benchmark and their use is so pervasive that generally they are regarded as fire demand regardless of the specifics.

Parametric fire curves represent a modest generalization of the fire curve approach (CEN [2002]). Typically, it is assumed that a building compartment is subjected to a uniform temperature distribution that follows the parametric fire curve. The curves include factors related to the compartment dimensions, size and number of openings, and amount of combustible materials and result in a unique intensity and duration for the fire. In general, parametric fire curves include a nonlinear heating phase followed by a linear cooling phase, while the standard fire is represented by an increasing curve (Figure 1). A further evolution of parametric fire curves is the use of ¿zone models¿ (Quintiere [1989]). In zone models the compartment is divided into multiple regions, each with its own uniform temperature distribution following a parameterized fire curve. Amongst other details, these models account for the fact that higher temperatures are observed in the upper zone of the compartment.

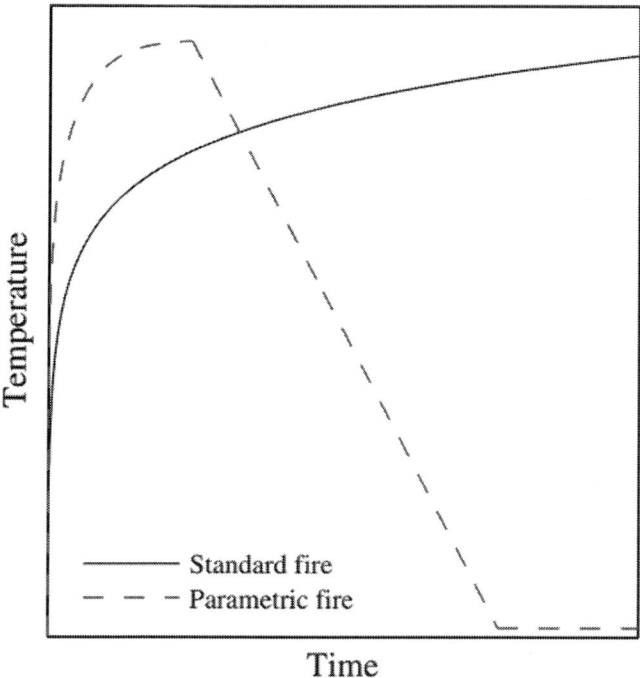

Figure 1: Fire curves.

The most sophisticated simulations adhere to computational fluid dynamics (CFD) and at some level attempt to model actual fire dynamics. These models are used to predict the development of fire in a building structure (including fully three-dimensional models), incorporating flames and smoke propagation. CFD simulations are complex, require a high level of expertise, relatively long computation times, and can be challenging to calibrate. Nonetheless, true performance-based design of fires relies on the long-term potential of this approach. Multiple software solutions are in current use, including PHOENICS (Spalding [1978]), FDS (McGrattan et al. [2002]), SMARTFIRE (Ewer et al.[2008]) and SOFIE (Rubini [2006]).

Heat Transfer

Once the thermal fire demands are established, the next step is to propagate these demands to the structure itself through heat transfer analysis. For an actual fire, temperature distributions on CFS members are non-uniform and vary through the cross-sections and along the length. Fully three-dimensional heat transfer models of CFS assemblages are possible (Santos et al. [2013]), but not common. Instead, simplified one-dimensional heat transfer models are generally used to estimate the temperature history across assemblages (walls, floors, etc.) subjected to fire action on one side (Sultan [1996]; Alfawakhiri [2001]; Keerthan and Mahendran [2013]; Chen et al. [2013]). These models require accurate thermal models of all materials in the assemblage. Generally, the heat conduction through and radiation from the steel studs are ignored, and shrinkage, cracking and ablation processes of the gypsum boards are not explicitly modeled, although their effects are considered by modifying material thermal conductivity and specific heat. Also, moisture migration and hot air flow are ignored, thus thermal gradients along the length of the assemblage (i.e., height of the wall) are ignored. It is also common to make the simplifying assumption that the temperature varies linearly throughout the web of the CFS member in the assemblage, while the temperature of the flanges and lips are constant (Shahbazian and Wang [2013]). See Performance of Walls section for further discussion of modeling heat transfer in CFS assemblages.

Cold-Formed Steel Material at Elevated Temperatures

During a fire, the temperature of structural members increases and, subsequently, material properties change. Mechanical properties of steel such as the elastic modulus, yield stress, and ultimate stress degrade with increasing temperature, thus steel members lose strength and stiffness under increasing temperature. Thermal properties also vary with temperature, e.g. steel incurs phase transformations under highly elevated temperatures that significantly alter thermal response. Quantification of the temperature dependence of thermal and mechanical properties of sheet steel is a fundamental building block for predicting the response of CFS under fire.

Mechanical Properties

Several research groups have studied mechanical properties of sheet steel at elevated temperatures (Lee et al. [2003]; Chen and Young [2007]; Ranawaka and Mahendran [2009a]; Kankanamge and Mahendran

[2011]; Chen and Ye [2012]). In general, tested specimens range from 0.50 mm [0.0188 in.] to 2.00 mm [0.0713 in.] thick, with yield strengths from 250 MPa [36 ksi] to 550 MPa [80 ksi] at ambient temperature. Typically, the experimental results are presented as retention factors, which are ratios of a material property at elevated temperature with respect to the same property at ambient conditions. Retention factors vary among research efforts (Figure 2) and the proposed prediction equations differ as well (Figure 3). Differences are mainly attributed to the test method, strain rate, heating rate, material grade, and the criteria used to determine the yield stress - as discussed further below.

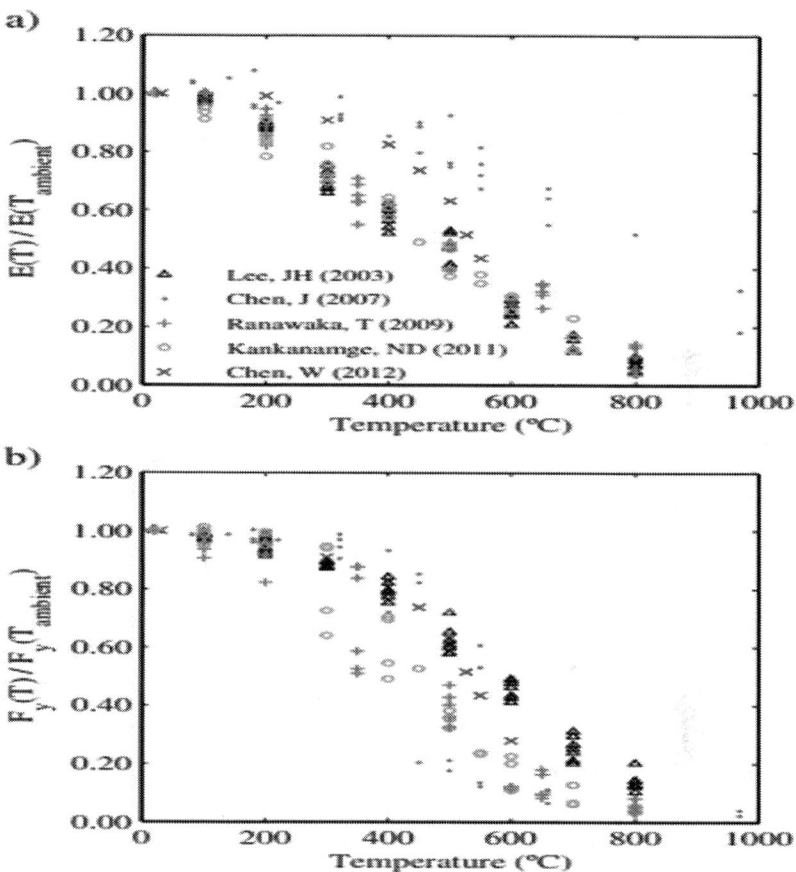

Figure 2: CFS retention factors for the (a) elastic modulus and (b) yield strength from steady-state tests.

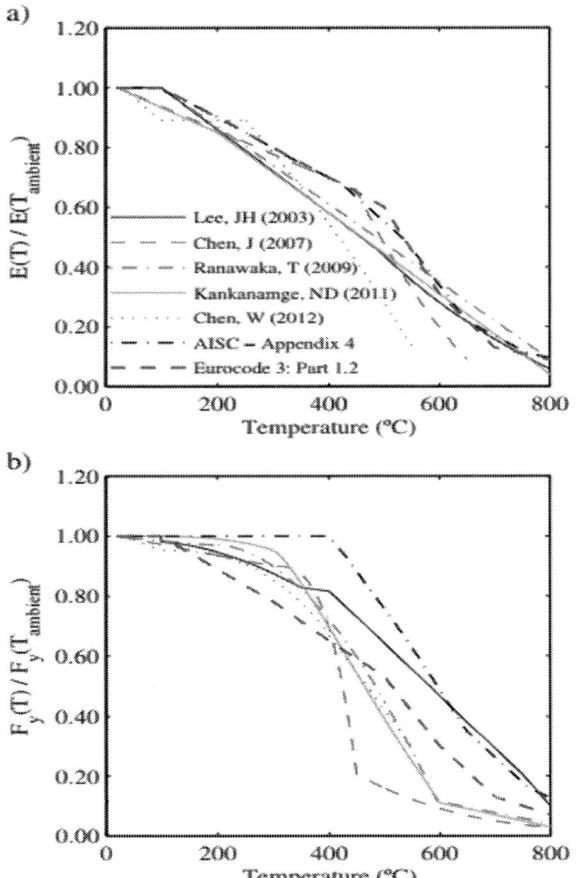

Figure 3: Proposed retention factors for the (a) elastic modulus and (b) yield strength of steel.

Tensile tests at elevated temperatures are traditionally conducted by either steady-state, or transient-state testing. During steady-state tests, the temperature on the specimen is increased to a given level and then, after the temperature becomes stable, external load is gradually applied until failure occurs. In contrast, during transient-state tests, the load is statically applied to the specimen, and the temperature is gradually increased until a failure criterion is met. Results are dependent on the test method. Although the steady-state test is more popular (and generally easier to conduct), the transient-state test is considered more realistic as it is consistent with a member under an applied static load (e.g. a gravity loaded column) undergoing temperature increase, as in a fire (Outinen and Mäkeläinen [1999]). In general, transient-state tests show a higher degradation than steady-state tests, e.g. see the retention for Young¿s modulus in Figure 4-a. Though common, the use of retention factors from steady-state tests may lead to overestimated stiffness and strength (Chen and Ye[2012]).

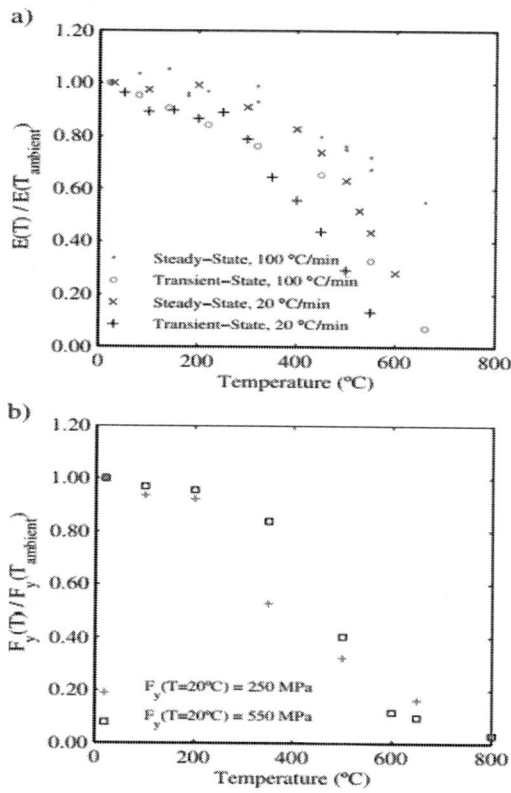

Figure 4: CFS retention factors for the (a) elastic modulus from steady state and transient state tests, and (b) yield strength for different steels.

Strain rate, typically not influential for sheet steel at ambient temperatures under common loading or testing rates, does influence the results in strain-controlled steady-state tests at temperature. Figure 2 shows that retention factors obtained using a strain rate of 0.006 min$^{?1}$ (Chen and Young[2007]) are higher than the factors obtained using strain rates at about 0.003 min$^{?1}$ (Lee et al.[2003]; Ranawaka and

Mahendran [2009a]; Kankanamge and Mahendran [2011]; Chen and Ye[2012]). In general higher strain rates lead to higher (stiffer) response (Cooke [1988]). In addition, retention factors for yield stress based on high strain rates often lead to a yield and ultimate strengths at similar magnitudes (Kankanamge and Mahendran [2011]). Thus, some care must be taken to insure strain rate is consistent with expected final use, when establishing retention factors.

High heating rates may also induce high strain rates during the heating process in transient-state tests (Outinen [2006]). Expected heating rates for structural steel members with 2 hour fire rated protection and unprotected sections are approximately 5.0?±?2.0°C/min and 32.5?±?7.5°C/min, respectively (Kodur et al. [2010]). Typically, heating rates adopted in transient-state tests on CFS specimens vary from 10 to 20°C/min. These values are not necessarily within the expected range during a fire, or even considered in computational simulations to predict the response of CFS structures. Figure 4a shows that retention factors obtained using a heating rate of 100°C/min (Chen and Young [2007]) are higher than the factors obtained using 20°C/min (Chen and Ye [2012]). High heating rates result in high-predicted strengths since material damage may be delayed under rapid temperature increases (Bednarek and Kamocka [2006]). Also, during tensile tests, the temperature is assumed to be uniform in the specimen, so it is important to provide enough time to stabilize the temperature and avoid significant thermal gradients. When the heating rate is high, it is more difficult to accurately monitor the temperature and guarantee a uniform distribution. In addition, heating rate alters the creep effect in sheet steel (Outinen [2006]).

It has been posited that differences in chemical composition lead to different retention factors as a function of steel grade (Ranawaka and Mahendran [2009a]). Young¿s modulus shows little dependence on steel grade, but retention factors for yield stress show a more complicated dependence. Researchers report that high strength steel (i.e. F_y?~?550 MPa) is more efficient than common (low strength) steel grades at about 400°C and above (Lee et al. [2003]). However, the available experimental data is limited, and mixed. As shown in Figure 4-b, the transition temperatures from high to low retention factors may be different depending on steel grade (Ranawaka and Mahendran [2009]a), but overall retention factors for yield stress are similar across grades. Determining yield stress retention factors is itself dependent on strain level, heat level, and the formal method for determining yield stress. The retention factors vary according to the yield point and yield strain definitions. Usually, the yield point is based on the 0.2% offset strain. However, other offset strains might be adopted to define the yield point. For instance, AISC ([2010]) provides yield stress retention factors for hot-rolled steel based on 2% offset strain. In general, the larger the offset strain used to define the yield point, the closer the retention factors for yield and ultimate stresses are.

Usually, temperature-dependent constitutive relations are based on the Ramberg and Osgood ([1943]) model:

$$\varepsilon_T = f_T/E_T + k_T \left(f_T/f_{y.T} \right)^{n_T}$$

(1)

Where ε_T, f_T, E_T and $f_{y.T}$ are the strain, stress, Young¿s modulus and yield stress at a temperature T (°C), respectively; and, k_T and n_T are parameters obtained from regression analysis. The Ramberg-Osgood strength coefficient, k_T, proposed by Chen and Young ([2006a]; [2007]) and Chen and Ye ([2012]) is 0.2%. According to temperature-dependent equations proposed by Mahendran and his colleagues, k_T ranges from 0.08% to 0.31% (Ranawaka and Mahendran [2009a]; Kankanamge and Mahendran [2011]); however, all cases the 0.2% offset method was used to compute the yield stress used in Ramberg-Osgood equations. At ambient temperature (around 25°C), typical Ramberg-Osgood hardening coefficients (n_T) for cold-formed stainless steels range from 4.5 to 12.2 (Rasmussen [2003]). However, n_T computed from temperature-dependent equations ranges from 17.2 (Chen and Young [2007]) to 57.6 (Ranawaka and Mahendran [2009a]) at ambient temperature, for cold-formed carbon steel G550, under steady-state testing conditions. According to proposed equations, n_T tends to decrease with increasing temperature, up to 450°C. Then, up to 800°C, n_T ranges from 4.6 (Chen and Young [2007]) to 24.8 (Chen and Ye [2012]). Clearly, more work is needed to clarify the correct application of Ramberg-Osgood expressions under temperature for low carbon sheet steels.

Current steel design codes (AISC [2010]; CEN [2005]) provide retention factors for mechanical properties of steel at elevated temperatures, as shown in Figure 3. General trends for CFS specimens are consistent with the wider database of tested hot-rolled steels, but material and test method dependent scatter exists and, in some cases, particularly around 400°C, observed reductions of yield stress are far greater in CFS than in the code-based expressions for hot-rolled steel.

The manufacturing process for CFS sections can create significant changes in the material properties, particularly near the corners, in a phenomenon typically referred to as cold work of forming. This additional cold work of forming strength is gradually lost with increasing temperatures (Mäkeläinen and Outinen [1998]; Lee et al. [2003]), and completely disappears above 500°C (SCI [1993]). However, CFS maintains its nominal yield strength without cold work of forming after heating and cooling (Outinen and Mäkeläinen [2004]).

Other material properties such as density and Poisson¿s ratio of steel are commonly assumed to be constant (Kaitila [2002]). Nevertheless, mass density slightly decreases (Costes [2004]) and Poisson¿s ratio increases (Clark [1953]) with increasing temperature (see Figure 5). Prediction equations for the Poisson¿s ratio of CFS are not available; however, working directly from the available data may be useful to infer other constitutive parameters, such as shear modulus.

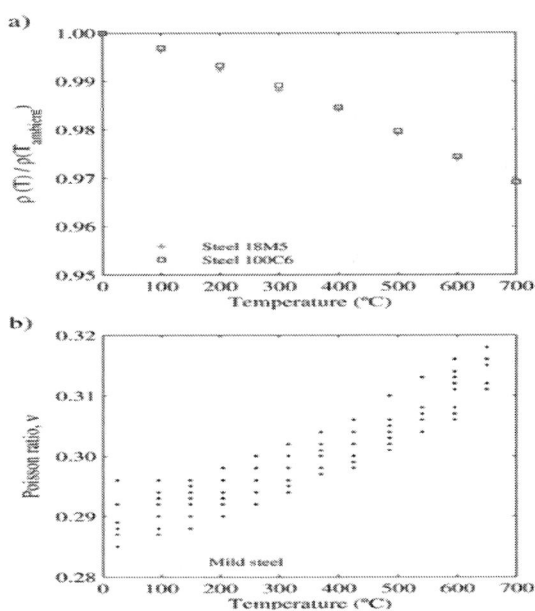

Figure 5: Temperature dependence of steel (a) density and (b) Poisson¿s ratio.

Significant limitations exist with the available data. At the most basic level, for use in the United States, the tested CFS does not conform to ASTM A1003 as specified in AISI ([2012a]). Further, the impact of temperature on residual stresses and strains has seen only limited study (Feng et al [2003a]; Lee et al. [2003]). Since the cold-working process influences both explicit design expressions (cold work of forming) and implicit design expressions (the basic column curve) the impact of temperature (potentially similar to annealing) could be influential. The relatively common practice of cold-reducing the steel to a desired thickness has also not seen separate study for its effect on properties under temperature. CFS creep effects and behavior after the cooling phase have been scarcely studied (Outinen [2006]). Moreover, unresolved issues at ambient temperature such as the difference in tensile and compressive yield strength in CFS (Karren [1970]; Uribe [1969]) also become more important as attempts to predict capacity are advanced.

In summary, research needs related to the mechanical properties include determining suitable heat rates that represent realistic fire conditions to study the material response during the heating and decaying phases of fire, and after cooling down. The influence of heat and load (or strain) rates on the mechanical properties at high temperatures needs to be studied, so the advantages and disadvantages of different types of test (i.e. steady-state and transient) are better comprehended. Furthermore, attention should be paid to the chemical composition of specimens tested since the mechanical response seems to differ among different materials, and even for the same material before and after the cold-reducing process. Since mechanical properties at elevated temperatures might depend on the loading conditions, compression tests are also needed for material characterization. Data needs also include the Poisson¿s ratio, shear modulus, and characteristic stress¿strain relations of CFS at elevated temperatures.

Thermal Properties

Thermal properties govern heat transfer and thermal deformations. Though important, they have seen less study than mechanical properties by the structural engineering community. At highly elevated temperature, steel may suffer a pearlite to austenite phase transformation, changing its internal crystal structure from body-centered cubic to face-centered cubic. During this transformation, no significant elongation (Chen and Ye [2012]) or contraction (Cooke [1988]) is observed. The temperature ranges at which these changes occur are sensitive to the chemical composition of the steel, but are generally high, and often higher than the temperatures at which structural failure is reached (Cooke [1988]).

As illustrated in Figure 6, at temperatures below the phase transformation in carbon steel: thermal strains grow nonlinearly with increasing temperature; heat capacity increases with increasing temperature; and, conductivity decreases with increasing temperature. As shown in Figure 6-a, it is common in some codified solutions (AISC [2010], AS [1998]), to ignore the temperature dependence of the thermal expansion coefficient. This should be done with some care, as the thermal expansion coefficient governs the thermal strain field of structural members and (depending on the displacement boundary conditions) controls the magnitude and shape of thermal deformations.

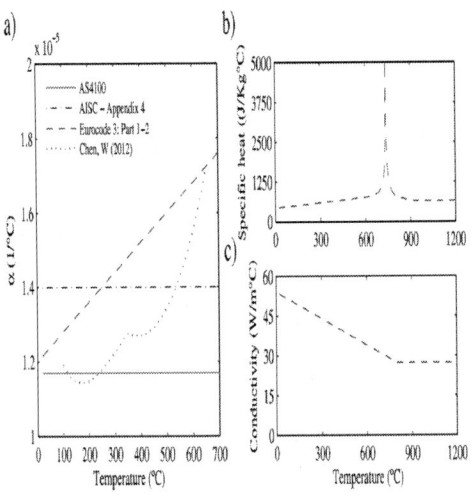

Figure 6: Thermal properties of steel at elevated temperatures (a) thermal expansion coefficient, (b) specific heat, and (c) thermal conductivity.

Research needs include material testing to characterize the thermal conductivity, specific heat and thermal expansion coefficient of CFS at elevated temperatures, as well as the identification of critical temperatures at which phase transformation occur, and the thermal and mechanical properties of the material are modified.

Cold-Formed Steel Members at Elevated Temperatures

The growing use of CFS in the construction industry has led to an increasing attention in the research community towards CFS performance under fire conditions. Thin-walled steel members are potentially more vulnerable to fire effects because of their high surface to volume ratio and relatively high thermal conductivity. If unprotected, these allow for rapid temperature increase, and consequently fast stiffness and strength degradation. Considering the temperature dependence of both mechanical and thermal properties, under realistic fire conditions, the stiffness, strength, and thermal elongation vary across the section of a member and along its length, creating a dynamically changing demand and capacity.

CFS members, as conventionally employed in light steel framing, are unique: efficient in terms of strength-to-weight, but markedly more complex than typical hot-rolled steel members due to their thin-walled nature and related cross-section stability modes that must be accounted for in design. Strength prediction of thin-walled CFS members relies either on the Effective Width Method (EWM) or the Direct Strength Method (DSM) to account for local and/or distortional buckling (see e.g., AISI[2012b]). DSM is preferred because it directly integrates (computational) elastic buckling analysis into the design process. This was originally envisioned as a means to handle the wide variety of different shapes that can be formed from sheet steel (Schafer [2006]), but can be modified to include the wide variety of different stiffness properties within a cross-section due to temperature gradients in the section.

Currently, the design of CFS members under fire is based on standard fire tests results, under controlled laboratory conditions. Fire resistance is judged based on the amount of time that a member or assembly can withstand elevated temperatures without exceeding specific failure criteria. This quantity is correlated with the amount of available time for occupant¿s evacuation and firefighter¿s operation before structural failure. To enable a more engineered solution, research studies are generally focused on predicting the load-carrying capacity of members at elevated temperatures, typically using modifications to existing design methods. This approach aims to use this strength prediction coupled with heat transfer analysis and a given fire demand to establish the building fire performance.

Columns

At ambient temperatures the capacity of a CFS column must consider the interaction of local, distortional, and global buckling as well as yielding. Under fire demand, all the buckling modes and yielding potentially become time and temperature dependent through the cross-section and along the length In addition, due to thermal elongation and shift in the center of resistance from the changing mechanical properties, second-order P-? demands driven by thermal deformations can be important (Wang and Davies [2000]).

At ambient temperature, certain modal interactions are generally considered (e.g., local¿global) while others disregarded (e.g. local-distortional). Under thermal gradients these interactions can become far more complex (Batista-Abreu and Schafer [2013]). For instance, experimental results qualitatively show that short columns with holes, dominated by local buckling at temperatures below 400°C, fail in distortional buckling at higher temperatures (Feng et al. [2003b]). Further, short columns without holes dominated by distortional mode at temperatures below 400°C exhibit local-distortional-global (flexural) interaction at higher temperatures (Feng et al. [2003b]). These evolutions of modal interaction, as a function of temperature, can be quantified through modal identification methods using the constrained finite strip method as a basis (Li et al. [2012]).

Experimental data shows that the axial capacity of columns is reduced with increasing temperatures (Feng et al. [2003b]). For instance, short columns develop substantial axial strength degradation after 200°C, withstanding about 15% of the failure load at ambient conditions at 700°C.

Computational mechanical models of CFS columns at elevated temperatures typically utilize shell finite elements, are only loosely coupled to thermal analyses, and focus on the impact of a uniform, elevated temperature on the collapse capacity of a CFS column (Feng et al. [2003c]; Kaitila [2002]; Ranawaka and Mahendran [2006]; Chen and Young [2006b]; Ranawaka and Mahendran [2009b]). The models use temperature dependent mechanical properties for E and F_y typically based on testing conducted by the authors or on available retention factors (e.g., CEN [2005]). Residual stresses are usually ignored (e.g. Ng and Gardner [2007]) as they tend to diminish with increasing temperature (Ranawaka and Mahendran [2006]; Lee [2004]) and their influence on measured compressive ultimate load is negligible (Ranawaka and Mahendran [2010]; Gardner and Nethercot [2004]; Ellobody and Young [2005]). In fact, this is consistent with findings at ambient temperatures as well (Schafer and Peköz [1998]; Schafer et al. [2010]).

Consistent with the thin-walled nature of the response, initial imperfections based on eigen- buckling modes are typically included in the models. At ambient temperatures significant progress has been made in realistic characterization of local, distortional, and global imperfections (Zeinoddini and Schafer [2012]); however, at elevated temperatures simpler approaches are typically employed for imperfection magnitudes: local?~?t, distortional ~2 t, global L/500, where t is the thickness and L the member length (Feng et al. [2004], Kaitila [2002]; Ranawaka and Mahendran [2010]). Under temperature gradients that are non-uniform through the cross-section, the necessity for fine-tuned imperfections is likely to be outweighed by the eccentricity in stiffness and the thermal bowing resulting from differential expansion (Feng and Wang [2005]).

Work has also been completed on design methods for CFS columns at elevated temperatures. Under uniform temperature the DSM formulation (AISI [2012b] Appendix 1) with updated E(T) and related elastic buckling loads, and $F_y(T)$ and related squash load have been used within the traditional DSM expressions with good success (Heva et al. [2008]; Ranawaka and Mahendran [2009b]). Thermal bowing is more pronounced under non-uniform temperature, and Shahbazian and Wang ([2011a],[2011b], [2012]) have proposed modified DSM expressions and a new approach to determining the squash load capacity. The results are sensitive to the variation in the temperature across the section: temperature ratios between the exposed and unexposed flanges of 3.0, 2.0 and 1.5 at 120 minutes under a standard fire curve are utilized. Experimental data indicates actual temperature ratios are time dependent (see Figure 7) leading to further complications and a necessity to more directly couple the thermal and mechanical/design response. These temperature ratios were obtained through thermocouples located on the outside surface of the corners of lipped channels.

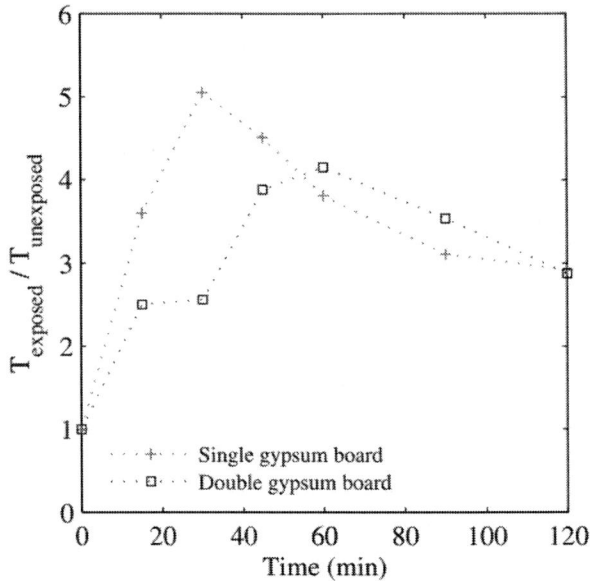

Figure 7: Temperature ratio between exposed and unexposed faces of a 100?×?54?×?15?×?1.2-lipped channel subjected to thermal load (cellulosic fire curve) on one flange (adapted from Feng et al.[2003a]).

Beams

Compared to columns, CFS beams under elevated temperature have seen relatively little study. Many of the challenges for columns are similar for beams: time-temperature dependence, altered buckling modes, modal interactions, and material yielding. Numerical investigations, based on shell finite element models, include work on lipped channels (Kankanamge and Mahendran [2008], [2012]) and zee shapes (Lu et al. [2010]). The models are subjected to uniform bending and analyzed with material properties consistent with uniformly elevated temperatures. Response is highly dependent on the end restraints, because they determine the development of compressive loads due to thermal elongation at initial stages of the fire action (Lu et al. [2011a]), and tensile forces due to catenary action during the fire response. Lateral restraint provided by sheathing is critical at ambient temperatures and under elevated temperatures. Prediction equations for lateral-torsional buckling of channel sections have been proposed (Kankanamge and Mahendran [2012]), but experimental and further numerical studies are needed.

In summary, existing research needs include the determination of realistic temperature distributions throughout the length and cross-section of CFS members. The structural response dynamically evolves as the temperature field changes; therefore, the study of mode interactions is important to understand the behavior of thin-walled members at elevated temperatures. Besides strength and stiffness degradation, structural members incur thermal deformations that would eventually lead to failure. Hence, the study of semi-rigid end restraints is crucial. In terms of experimental data, very limited information is available on the behavior of single sections other than channels, and composite sections. Furthermore, design methods such as the Direct Strength Method and Effective Width Method have been validated for limited high temperature conditions; however this validation does not represent a sufficiently extensive range of possible scenarios.

Cold-Formed Steel Connections at Elevated Temperatures

Connections are critical in understanding the performance of CFS structures under elevated temperatures. At ambient temperatures a significant body of literature exists on bolted steel-to-steel connections, with more limited studies on other fasteners and sheet steel connected to other materials (wood products, gypsum products, concrete, etc.). The knowledge base is similar, but with less depth of results, for connection performance at elevated temperatures.

Young and his colleagues have studied bolted steel-to-steel connections relevant to CFS construction at elevated temperatures, including proposed reduction factors (Lim and Young [2007]), extensive (120 specimens) steady-state tests and analysis on single shear bolted connections (Yan and Young[2011a], Yan and Young [2012a]), and complementary (62 specimen) transient-state tests (Yan and Young [2011b]). The tests and analysis show the dominance of bearing failures as long as the ¿3d¿ edge distance criteria is maintained, and also show that the use of reduced mechanical properties at elevated temperatures, but traditional ambient temperature strength equations, provides an adequate prediction of strength. Results show the capacity of connections is significantly reduced with increasing temperature. For instance, experimental data shows a degradation of the bearing strength of bolted moment connections up to 90% at 700°C, with respect to its capacity at ambient conditions.

Tests on screw fastened steel-to-steel connections in single shear under steady-state (Yan and Young[2012b]) and transient-state (Yan and Young [2012c]) conditions lead to similar findings as bolted connections. In addition, Lu et al. ([2012]) numerically studied shot-nailed and screwed connections, and again found that bearing failure of the thin steel sheet was the dominant failure mode. Design guidelines were provided to predict the capacity of shot-nailed (Lu et al. [2013]), and screwed connections at elevated temperatures (Lu et al. [2011b]).

In general, sheathed members are more stable and develop higher load-carrying capacity than unsheathed members. A methodology for sheathing-braced design of studs based on experimental data and discrete spring models is utilized at ambient conditions. However, studies on stud-to-sheathing connections at high temperatures are not currently found in the literature. Therefore, the feasibility of a similar methodology for fire design of CFS structures has not been judged. Research needs include the study of steel stud-to-sheathing connections at high temperatures. Additionally, the heat transfer through steel-to-steel and steel stud-to sheathing connections is relevant to understand the global behavior of CFS systems under fire.

Cold-Formed Steel Assemblages at Elevated Temperatures

While CFS material and member performance under elevated temperature represents important building blocks for understanding fire resistance, it is complete CFS assemblages (i.e. walls and floors) that provide structural support and resist fire demands. The standard approach for assessing walls and floors is the performance in a standard fire test, as discussed in detail below. Industry has performed such testing extensively for CFS framing assemblages (CFSEI [2012]). From the standpoint of the development of performance-based design, these tests provide benchmarks that the development of analysis-based approaches may be compared with. Thus, understanding the standard fire test and response of CFS assemblages is an important step in understanding full fire response, but must be coupled with more advanced fire demand and heat transfer models to provide a complete prediction of response.

Standard Fire Testing

The Standard Test Methods for Fire Tests of Building Construction and Materials (ASTM [2012]) are the most commonly referenced methods for fire testing of CFS assemblages. Equivalent, or similar, test standards also exist (UL [2003]; ISO [1999]). The fire curve used in ASTM was developed in 1918 (Manzello et al. [2008a]), is equivalent to UL 263, and has a higher initial rate of temperature rise compared to the ISO 834 (ISO [1999]) fire curve. Thus, for short test durations, the ASTM fire curve is more severe. However, fire curves have been strongly criticized due to the difference found between standard curves and fire curves measured in real compartment fires, both in terms of severity and duration. In this sense, the fire resistance specified for an assembly through standard testing may be different from the real response of the structure (Lane [2000]). The worth of the standard test is more in its comparison to past practice, that in its absolute response.

A standard fire test is illustrated in Figure 8. The specimens (wall or floors) are subjected to a specific and prescriptive time-temperature curve (Figure 8-a). Thermocouples are strategically located on the specimens and they are monitored throughout the test (Figure 8-b and c). Fire resistance is defined by the time until ¿the maximum temperature increase on the unexposed side of the wall exceeds 181°C (325°F); the average temperature increase on the unexposed side of the wall exceeds 139°C (250°F); a breach occurs in the wall that allows hot gases from the furnace to penetrate and ignite a cotton target on the unexposed side of the wall; or, the wall is unable to maintain its design load.¿ (ASTM [2012]). Test setup and response for typical tests are provided in Figure 8d-f for a wall and Figure 8g-i for a floor.

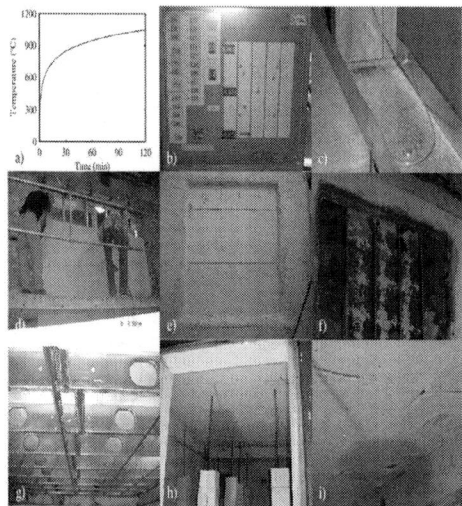

Figure 8: Proprietary ASTM fire test results on walls (d-f) and floors (g-i) provided by ClarkDietrich (a) prescriptive time-temperature curve, (b) thermocouple readings from wall test, (c) thermocouple installation, (d) installation of wall in furnace, (e) wall exterior during test, (f) wall interior and calcined gypsum board after test, (g) proprietary floor system showing blocking and strapping of joists (h) underneath floor before test, (i) after test.

Criticisms of the standard fire test are well summarized by Grosshandler ([2007]): ¿The maximum size of the wall system is limited by the size of the furnace. The load conditions for the test specimen may not adequately mimic field use. The thermal environment of the furnace does not mimic a real fire. The tests reveal no fundamental information about the performance of the specimen and provide little guidance on how to improve performance. The furnaces themselves are not standardized; hence, the same specimen could receive different ratings if tested in two different facilities. Ratings are based upon a single test, with no way to quantify the uncertainty or safety factor¿. In many ways the fact that ¿the tests reveal no fundamental information about the performance of the specimen¿ is the most damning and demonstrates how current practice provides no path towards significant improvement or change when driven by the standard fire test.

Even to use standard fire testing to advance basic modeling can be challenging since little, if any, of the specific data (thermocouple readings, deformations, etc.) is available in the public domain. Further, the pass/fail nature of the test has precluded studies focused on better understanding behavior. For example, the interaction between damage due to structural loads and degrading strength under realistic fire conditions has not been studied in detail.

By analyzing the data generated from standard fire tests, Ingberg (1928) developed a method to approximate the fire resistance time of a structure under a real fire, based on the fire resistance of a structure under standard fire conditions. This methodology compares the severities of real and standard fires, quantified as the areas under both fire curves. Other methods attempt to estimate ¿real¿ fire resistance rating based on the maximum temperatures that structural members develop. In general, these methodologies do not explicitly account for factors such as the type of combustible, geometry of the compartment, ventilation conditions, and heat release rate. Equations used to estimate the fire resistance rating are based on regression of limited experimental data related to tests with specific configurations and materials. Additionally, these methodologies do not consider the effect of loading conditions on the structural members, and the variation of temperature throughout the compartment.

Performance of Walls

Performance of CFS walls in standard fire testing is summarized in CFSEI ([2012]). For both partition walls and for load bearing walls, the fire rating is largely a function of the thickness and number of gypsum (or similar) wallboards. Thus, the primary interest in research has been on the heat transfer aspects of the gypsum wallboard under the standard fire curve. The role of the fasteners as a thermal bridge, the role of the cross-section stiffness with respect to thermal bowing of the wall and local flange deformations, and the role of lost axial capacity due to decreased bracing stiffness from the wallboard as the board burns and undergoes calcination (or is saturated by a sprinkler) are important, but have seen little or no study.

The performance of the wallboard itself directly drives the thermal response and indirectly influences the mechanical response of the system. Wallboards consist of a pressed gypsum (and glass fiber) core, covered with thick sheets of paper. The sheets of paper maintain the integrity of the gypsum core even when it cracks, until they burn at about 200°C to 300°C. Dehydration of gypsum plasterboards initiates at 100°C (Gerlich [1995]; Ngu [2004]), when water boils, leading to increased porosity and a considerable drop of thermal conductivity (Rahmanian [2011]). For instance, calcination of the gypsum board is complete after 20 minutes at 400°C, resulting in ~20% density reduction and ~80% loss of material strength (Cramer et al. [2003]), as shown in Figure 9. Gypsum board damage depends on the maximum temperature reached and the rate of temperature increase and its relations to the moisture flow, ablation and cracking processes (Ariyanayagam and Mahendran[2012]). Alternatives to gypsum wallboards such as bolivian magnesium and calcium silicate boards have shown better fire resistances (Chen et al. [2012]), but are associated with increased cost.

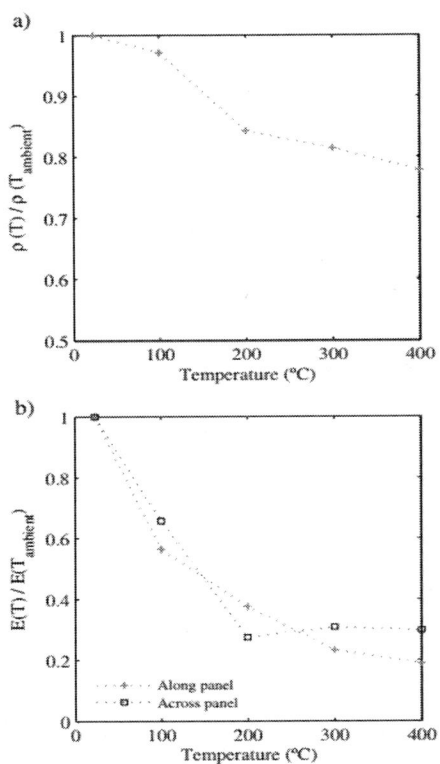

Figure 9: (a) Density and (b) elastic modulus retention of type X gypsum board (adapted form Cramer et al.[2003]).

Available data on thermo-mechanical properties of gypsum boards is commonly derived from research on lightweight wood construction. Thermal properties (i.e. specific heat, thermal conductivity, contraction and mass loss) of types X, C, F and R gypsum boards are available in the literature (Bakhtiary et al. [2000]; Bénichou and Sultan [2005]; Manzello et al. [2008a]; Manzello et al.[2008b]; Thomas [2002]). Variability in the chemical composition and testing conditions (e.g. heating rate) of gypsum leads to scattered thermal properties results (Wakili and Hugi [2009]). However, the chemical composition of the tested gypsum boards is not commonly stated in experimental reports. Modified thermal properties to implicitly account for mass transfer (e.g. water migration and re-condensation) and ablation process have also been proposed (Ang and Wang [2004]; Rahmanian[2011]).

Data on the mechanical properties of gypsum boards is scarce. Fuller ([1990]) showed the stiffness and strength of gypsum boards decays by 80% at 120°C, after calcination. Similarly, Cramer et al. ([2003]) reported the variation of the elastic modulus, bending strength and thermal expansion coefficient along and across type X gypsum boards, up to 400°C, after 60 minutes of fire exposure. Furthermore, Rahmanian ([2011]) reported the elastic modulus, bending and compressive strengths, and proposed linear stress¿strain relations for standard and glass-fiber reinforced gypsum boards at elevated temperatures.

CFS walls commonly contain cavity insulation for climate and sound control. In general, cavity insulation obstructs heat dissipation in the cavity, causing a faster temperature increase in the exposed face of the wall, while delaying the temperature increase in the unexposed face (Alfawakhiri and Sultan [2001]). Whether or not this change in the heat transfer is beneficial or detrimental is an open question. Research indicates glass or cellulose fiber cavity insulation has little affect, but mineral fiber insulation generally increases fire resistance (Sultan and Lougheed [2002], Feng et al.[2003d]). However, others conclude cavity insulation is generally detrimental to strength (Kolarkar[2010]). Alternatives to cavity insulation have been explored by Mahendran and his colleagues, including glass and rock fiber external insulation and external insulation sandwiched between gypsum boards. Fire performance for these systems can be excellent (Kolarkar and Mahendran[2008], [2012]; Gunalan and Mahendran [2010]; Keerthan and Mahendran [2012], [2013]).

Performance of Floors

CFSEI ([2012]) summarizes available sources for prescriptive fire design of CFS floor systems; however, only limited information on the behavior of CFS floor-ceiling systems is available. Sultan et al. ([1998]) tested five floor specimens with CFS joists and gypsum board sheathing under standard fire conditions. It was observed that thermal bowing of the steel joists governs floor deflections until run-away occurs. In general, local buckling at the top flange of the joists near mid-span, and subsequent inelastic mechanism formation led to structural failure. Conclusions indicate that cavity insulation has a detrimental effect on the fire resistance of floor systems especially when the insulation melts, allowing the CFS joists to be completely exposed to fire after the gypsum falls off. The ability of the sheathing boards to remain in place governs the overall fire resistance of CFS floor assemblies (Alfawakhiri and Sultan [2001]).

Baleshan and Mahendran ([2010]) tested three floor-ceiling systems looking at the advantages of using sandwiched insulation on the ceiling side of the CFS frame instead of cavity insulation. Results demonstrate that sandwiched insulation improves the fire resistance of floor systems by mitigating convective and radiative heat transfer from the external heat source to the CFS joists. It was observed that gypsum boards prevented lateral-torsional buckling of the joists during the tests. At high temperatures, local buckling along the CFS joist was prevalent, and pronounced crippling occurred near the supports.

The list of research needs related to CFS assemblages is exhaustive, as detailed in the previous sections. These needs begin with the characterization of fire demands based on realistic fire scenarios. Then, the actual heat transfer throughout the elements of the assemblage should be understood. The development of more accurate tridimensional and simplified heat transfer models is essential to enable coupled thermo-mechanical models useful for engineering-based analysis method. Furthermore, the study of the degradation of the capacity of the assemblage itself, and its interaction with the entire system is necessary for the development of a performance-based design method.

The Case for Performance-Based Fire Design of Cold-Formed Steel Systems

Fire represents one of the most important hazards that building structures must be designed against. Most of the modern regulatory framework around building structures originated in response to the great fires of the late 19th and early 20th century. Today, this regulatory framework supports material standards that have adopted reliability-based design methods that largely bring other building hazards (snow, wind, conventional dead and live load) into a risk consistent framework. For complex hazards such as earthquakes the risk consistent methodology has been extended to cover multiple performance objectives, always insuring society¿s concern of minimizing life safety risk, but enabling and incentivizing engineers and building owners to consider higher levels of performance. This performance-based design approach is largely seen as the future, and provides the best potential for risk consistent multi-hazard design.

Fire resistance of buildings framed from CFS is guaranteed through prescriptive codes and the standardized test. While the level of safety has generally been found acceptable, the lack of an engineering/ analysis-based approach to fire resistance of CFS structures impedes progress and stifles innovation, summarized here across four broad categories. First, the cost to industry, particularly for ASTM testing, is high and as a result little improvements are sought or found in even basic CFS wall and floor designs. Second, system-level mechanisms that provide enhanced resistance to fire through re-distribution of load are neither conceptualized, nor tested, nor designed in CFS structures due to lack of knowledge to complete such an approach and lack of financial reward for the engineer to do so. Third, risk consistent multi-hazard based design with fire is largely impossible since fire cannot be reasonably integrated with other hazards without a means to analyze the structure. Fourth, and finally, as multiple parties work to re-envision buildings to be greener and more sustainable, the current prescriptive approach to fire means fire protection is added as a constraint with a small set of known solutions instead of integrated within the larger optimization that needs to be performed.

Preliminary work on performance-based fire resistance has begun internationally, but it is still limited. Positive strides in this direction include codification of engineering/analysis-based methods that predict the response of structures under fire demands such as those recently adopted for hot-rolled steel (AISC [2010]). However, similar progress has not been made in this direction for CFS structures. CFS structures provide a compelling and challenging framework for advancing performance-based fire resistance.Compelling, because a significant percentage of the modern building stock uses CFS framed walls with gypsum board for interior fire resistance; when these same walls are load bearing, as in a highly efficient CFS framed building, additional considerations arise. Challenging, because the thin-walled nature of CFS members complicates conventional design significantly and at elevated temperatures the stability response is further modified and must be understood.

Significant challenges remain to developing a complete performance-based fire engineering solution for CFS structures; including: more realistic fire models; deeper understanding of the temperature dependence of CFS, gypsum, and connector thermal properties; three dimensional heat transfer models including gypsum board deterioration (dehydration, cracking, and ablation) processes; improved one-dimensional heat transfer models for design; better understanding of strength and stiffness degradation of CFS and connected wallboards at elevated temperatures, considering the influence of the test method, strain and heating rates and chemical composition; verified coupled thermo-mechanical models that accurately predict the response of CFS elements and subsystems; experimental research at large scale on CFS building structures, including fire development, cooling phase behavior and residual strength, and element and subsystems interactions; performance-based methods for CFS fire design; structural optimization of load-bearing and non-load-bearing fire protection systems; and multi-hazard building response and mitigation, including fire after earthquake and fire after blast.

CONCLUSIONS

Although work remains, the basic building blocks for analysis-based fire resistance of cold-formed steel (CFS) building assemblages and structures are in place and performance-based fire design for CFS structures can now be pursued. Prescriptive solutions under standard fire testing provide a variety of immediately available options for design, but are restricting innovation in CFS assemblages and systems and ultimately place fire outside of the risk-consistent framework that has been developed for other building natural hazards. Ideally, performance-based fire design brings the demand (fire modeling), propagation (heat transfer), and capacity (strength at elevated temperatures) all into the realm of analysis. This review article briefly summarizes current efforts in fire modeling and heat transfer. For capacity determination, the paper provides detailed reviews and composite data on the temperature dependence of sheet steels commonly used in CFS, members formed from CFS, and wall and floor assemblages framed from CFS. Available data is compared, along with existing codified provisions for other steels, and recommendations are provided for modeling and expected performance whenever possible. Codified provisions for analysis-based fire design of CFS will enable performance-based fire engineering of CFS structures and should be a near term goal. Work remains to provide detailed capacity predictions fully coupled with three-dimensional building models and simulated fires, but already this possibility exists for researchers, and in the future for designers as well.

AUTHORS¿ CONTRIBUTIONS

All listed authors made substantial intellectual contribution to this work; and read and approved the final manuscript.

ACKNOWLEDGEMENTS

The authors would like to thank ClarkDietrich for providing pictures taken during fire testing of wall and floor assemblages.

REFERENCES

1. Alfawakhiri F (2001) Behaviour of cold-formed-steel-framed walls and floors in standard fire resistance tests. Carleton University, Ottawa, ON.

2.   Alfawakhiri F, Sultan MA (2001) Numerical modelling of steel members subjected to severe thermal loads. National Research Council Canada, Ottawa, ON.

3.   Allen D (2004) Designing Cold-Formed Steel Mid-Rise Structures: Innovations for Cost-Effective and Attractive Projects. Structure Magazine, Chicago, IL.

4.   (2010) Specification for Structural Steel Buildings, Appendix 4: Structural Design for Fire Conditions. American Institute of Steel Construction, Chicago, IL.

5.   (2012a) AISI S200-12: North American Standard for Cold-Formed Steel Framing ¿ General Provisions. American Iron and Steel Institute, Washington, DC.

6.   (2012b) AISI S100-12: North American Standard for the Design of Cold-Formed Steel Structural Members. American Iron and Steel Institute, Washington, DC.

7.   (2012) ASTM E119-12a: Standard Methods of Fire Tests for Fire Tests of Building Construction and Materials. American Society for Testing and Materials, West Conshohocken, PA.

8.   (2009) Performance-Based Design of Structural Steel for Fire Conditions: A Calculation Methodology. American Society of Civil Engineers, Reston, VA.

9.   Ang CN, Wang YC (2004) the effect of water movement on specific heat of gypsum plasterboard in heat transfer analysis under natural fire exposure. Construction and Building Materials 18(7):505-515 [http://dx.doi.org/10.1016/j.conbuildmat.2004.04.003]   http://dx.doi.org/10.1016/j.conbuildmat.2004.04.003.

10.   Ariyanayagam AD, Mahendran M (2012) Fire Tests of Load Bearing Steel Stud Walls Exposed to Real Building Fires. Proceedings of the 7th International Conference on Structures in Fire. ETH Zürich, Zürich, Switzerland.

11.   Bakhtiary SB, Jafarpoor F, Firoozyar F (2000) Thermal and mechanical properties of fire-resistant gypsum plasters. Asian J Civ Eng (Building and Housing) 1(2):67-82

12.   Baleshan B, Mahendran M (2010) Full Scale Fire Tests of A New Light Gauge Steel Floor-Ceiling System, Proceedings of the 4th International Conference on Steel & Composite Structures. Queensland University of Technology, Sydney, Australia, 21¿23 July 2010.

13. Batista-Abreu JC, Schafer BW (2013) Stability Of Cold-Formed Steel Compression Members Under Thermal Gradients. Proceedings of the Annual Stability Conference Structural Stability Research Council. In: Missouri University of Science and Technology, Rolla, MO, pp 136¿154, 17-19 April 2013.

14. Bednarek Z, Kamocka R (2006) The heating rate impact on parameters characteristic of steel behaviour under fire conditions. J Civ Eng Manag 12(4):269-275 doi:10.1080/13923730.2006.96 36403.

15. Bénichou N, Sultan MA (2005) Thermal properties of lightweight-framed construction components at elevated temperatures. Fire and Materials 29(3):165-179 doi:10.1002/fam.880.

16. (2012) Technical Note on Cold-Formed Steel Construction, TECH-NOTE T100-12: Fire Rated Assemblies of Cold-Formed Steel Construction. Cold-Formed Steel Engineers Institute, Washington, DC.

17. Chen W, Ye J (2012) Mechanical properties of G550 cold-formed steel under transient and steady state conditions. J Constructional Steel Res 73:1-11

18. Chen J, Young B (2006) Corner properties of cold-formed steel sections at elevated temperatures. Thin-Walled Structures 44(2):216-223 doi:10.1016/j.tws.2006.01.004.

19. Chen J, Young B (2006b) Design of cold-formed steel lipped channel columns at elevated temperatures, Proceedings of the International Colloquium on Stability and Ductility of Steel Structures. Instituto Superior Técnico, Lisbon, Portugal.

20. Chen J, Young B (2007) Experimental investigation of cold-formed steel material at elevated temperatures. Thin-Walled Structures 45(1):96-110 doi:10.1016/j.tws.2006.11.003.

21. Chen W, Ye J, Bai Y, Zhao X (2012) Full-scale fire experiments on load-bearing cold-formed steel walls lined with different panels. J Constructional Steel Res 79:242-254

22. Chen W, Ye J, Bai Y, Zhao X (2013) Thermal and mechanical modeling of load-bearing cold-formed steel wall systems in fire. J Structural Eng doi:10.1061/(ASCE)ST.1943-541X.0000862.

23. Clark CL (1953) High-temperature alloys. Pitman Metallurgy Series. Pitman Publishing Corporation, Toronto.

24. Cooke GME (1988) An introduction to the mechanical properties of structural steel at elevated temperatures. Fire Safety J 13(1):45-54

25. Costes F (2004) Modélisation thermomécanique tridimensionnelle par éléments finis de la coulée continue d¿aciers. Ecole Nationale Supérieure des Mines de Paris, Paris, France.

26. Cramer SM, Friday OM, White RH, Sriprutkiat G (2003) Mechanical Properties Of Gypsum Board At Elevated Temperatures. In: Fire And Materials 2003: 8th International Conference. Interscience Communications Limited, San Francisco, CA, USA. pp 33-42

27. Ellobody E, Young B (2005) Structural performance of cold-formed high strength stainless steel columns. J Constructional Steel Res 61(12):1631-1649

28. (2002) BS EN 1991-1-2:2002 Eurocode 1: Actions on Structures Part 1¿2: General actions ¿ Actions on structures exposed to fire. European Committee for Standardization, Brussels, Belgium.

29. (2005) BS EN 1993-1-2:2005 Eurocode 3: Design of Steel Structures. Part 1¿2: General Rules - Structural Fire Design. European Committee for Standardization, Brussels, Belgium.

30. Ewer J, Jia F, Grandison A, Galea E, Patel M (2008) SMARTFIRE V4.1: User Guide and Technical Manual - SMARTFIRE Tutorials. University of Greenwich, London, United Kingdom.

31. Feng M, Wang YC (2005) An analysis of the structural behaviour of axially loaded full-scale cold-formed thin-walled steel structural panels tested under fire conditions. Thin-Walled Structures 43(2):291-332

32. Feng M, Wang YC, Davies JM (2003) Axial strength of cold-formed thin-walled steel channels under non-uniform temperatures in fire. Fire Safety J 38(8):679-707 doi:10.1016/s0379-7112(03)00070-5.

33. Feng M, Wang YC, Davies JM (2003) Structural behaviour of cold-formed thin-walled short steel channel columns at elevated temperatures. Part 1: experiments. Thin-Walled Structures 41(6):543-570 doi:10.1016/s0263-8231(03)00002-8

34. Feng M, Wang YC, Davies JM (2003) Structural behaviour of cold-formed thin-walled short steel channel columns at elevated

temperatures. Part 2: Design calculations and numerical analysis. Thin-Walled Structures 41(6):571-594 doi:10.1016/s0263-8231(03)00003-x

35. Feng M, Wang YC, Davies JM (2003) Thermal performance of cold-formed thin-walled steel panel systems in fire. Fire Safety J 38(4):365-394 doi:10.1016/s0379-7112(02)00090-5

36. Feng M, Wang YC, Davies JM (2004) A numerical imperfection sensitivity study of cold-formed thin-walled tubular steel columns at uniform elevated temperatures. Thin-Walled Structures 42(4):533-555

37. Fuller JJ (1990) Predicting the Thermo-Mechanical Behavior of A Gypsum-To-Wood Nailed Connection. Corvallis, OR, Oregon State University.

38. Gardner L, Nethercot D (2004) Numerical Modeling of Stainless Steel Structural Components¿A Consistent Approach. J Structural Eng 130(10):1586-1601 doi:10.1061/(ASCE)0733-9445(2004)130:10(1586)

39. Gerlich JT (1995) Design of Loadbearing Light Steel Frame Walls for Fire Resistance. vol Fire Engineering Research Report 95/3. School of Engineering, University of Canterbury, Christchurch, New Zealand.

40. Grosshandler WL (2007) Fire Resistance Proficiency Testing of Gypsum/Steel-Stud Wall Assemblies, Technical Memorandum of PWRI 4075, Wind and Seismic Effects, U.S./Japan Natural Resources Development Program (UJNR), Joint Meeting, 39th. Public Works Research Institute, Tsukuba, Japan.

41. Gunalan S, Mahendran M (2010) Structural and Fire Behaviour of a New Light Gauge Steel Wall System, Proceedings of the 6th International Conference on Structures in Fire. DEStech Publications, Inc, Lancaster, PA.

42. Heva B, Yasintha D, Mahendran M (2008) Local Buckling Tests of Cold-formed Steel Compression Members at Elevated Temperatures. Proceedings of the Paper presented at the 5th International Conference on Thin-walled Structures - ICTWS. Innovations in Thin-walled Structures, Gold Coast, Australia.

43. (2012) International Building Code Chapter 6: Types of Construction. International Code Council, Country Club Hills, IL.

44. (1999) ISO 834¿1:1999: Fire Resistance Tests ¿ Elements of Building Construction. International Organization for Standardization, Geneva, Switzerland.

45. Kaitila O (2002) Finite element modelling of cold-formed steel members at high temperatures. Helsinki University of Technology, Espoo, Finland.

46. Kankanamge ND, Mahendran M (2008) Numerical Studies Of Cold-Formed Steel Beams Subject To Lateral-Torsional Torsional Buckling At Elevated Temperatures, Proceedings of the 5th International Conference on Thin-Walled Structures. Queensland University of Technology, Brisbane, Australia.

47. Kankanamge ND, Mahendran M (2011) Mechanical properties of cold-formed steels at elevated temperatures. Thin-Walled Structures 49(1):26-44 doi:10.1016/j.tws.2010.08.004

48. Kankanamge ND, Mahendran M (2012) Behaviour and design of cold-formed steel beams subject to lateral¿torsional buckling at elevated temperatures. Thin-Walled Structures 61:213-228

49. Karren KW (1970) Corner Properties Of Cold-Formed Steel Structural Members. Effects of Cold Work In Cold-Formed Steel Structural Members. Cornell Engineering Research Bulletin, Ithaca, NY.

50. Keerthan P, Mahendran M (2012) Numerical Modelling of Load Bearing LSF Walls Under Fire Conditions, Proceedings of the 7th International Conference on Structures in Fire. ETH Zürich, Zürich, Switzerland.

51. Keerthan P, Mahendran M (2013) Thermal performance of composite panels under fire conditions using numerical studies: plasterboards, rockwool, glass fibre and cellulose insulations. Fire Technol 49(2):329-356 doi:10.1007/s10694-012-0269-6

52. Kodur V, Dwaikat M, Fike R (2010) High-temperature properties of steel for fire resistance modeling of structures. J Mater Civ Eng 22(5):423-434 doi:10.1061/(asce)mt.1943-5533.0000041

53. Kolarkar PN (2010) Structural and Thermal Performance of Cold-formed Steel Stud Wall Systems under Fire Conditions. Queensland University of Technology, Brisbane, Australia.

54. Kolarkar P, Mahendran M (2008) Thermal Performance of Plasterboard Lined Steel Stud Walls, Proceedings of the 19th

International Specialty Conference on Cold-Formed Steel Structures. Missouri University of Science and Technology, Rolla, MO.

55. Kolarkar P, Mahendran M (2012) Experimental studies of non-load bearing steel wall systems under fire conditions. Fire Safety J 53:85-104

56. Lane B (2000) Performance-Based Design for Fire Resistance, Modern Steel Construction. American Institute of Steel Construction, Chicago, IL.

57. Lee JH (2004) Local buckling behaviour and design of cold-formed steel compression members at elevated temperatures. Queensland University of Technology, Brisbane, Australia.

58. Lee JH, Mahendran M, Makelainen P (2003) Prediction of mechanical properties of light gauge steels at elevated temperatures. J Constructional Steel Res 59(12):1517-1532 doi:10.1016/s0143-974x(03)00087-7

59. Li Z, Batista Abreu JC, Adany S, Schafer BW (2012) Cold-formed steel member stability and the constrained finite strip method. Proceedings of the 6th International Conference on Coupled Instabilities in Metal Structures. CIMS, Glasgow, United Kingdom.

60. Lim JBP, Young B (2007) Effects of elevated temperatures on bolted moment-connections between cold-formed steel members. Eng Struct 29(10):2419-2427

61. Lu W, Makelainen P, Outinen J (2010) Numerical investigation of cold-formed steel purlin in fire. Journal of Structural Mechanics 43(1):12-24

62. Lu W, Makelainen P, Outinen J (2011a) Behaviour of Cold-Formed Z-Shaped Steel Purlin in Fire. Proceedings of Paper presented at the Civil Engineering ¿11 - 3rd International Scientific Conference. Jelgava, Latvia.

63. Lu W, Mäkeläinen P, Outinen J, Ma Z (2011) Design of screwed steel sheeting connection at ambient and elevated temperatures. Thin-Walled Structures 49(12):1526-1533

64. Lu W, Ma Z, Mäkeläinen P, Outinen J (2012) Behaviour of shear connectors in cold-formed steel sheeting at ambient and elevated temperatures. Thin-Walled Structures 61:229-238

65. Lu W, Ma Z, Mäkeläinen P, Outinen J (2013) Design of shot nailed steel sheeting connection at ambient and elevated temperatures. Eng Struct 49:963-972

66. Mäkeläinen P, Outinen J (1998) Mechanical properties of an austenitic stainless steel at elevated temperatures. J Constructional Steel Res 46(1¿3):455

67. Manzello SL, Grosshandler WL, Mizukami T (2008a) Furnace Testing of Full-Scale Gypsum Steel Stud Non-Load Bearing Wall Assemblies: Results of Multi-Laboratory Testing in Canada, Japan and USA, Proceedings of the 5th International Conference on Structures in Fire. Research Publishing Services, Singapore.

68. Manzello SL, Park SH, Mizukami T, Bentz DP (2008b) Measurement Of Thermal Properties Of Gypsum Board At Elevated Temperatures. Proceedings of the 5th International Conference on Structures in Fire. Nanyang Technological University, Singapore.

69. McGrattan KB, Floyd JE, Hostikka S, Prasad KR (2002) Fire Dynamics Simulator (Version 3): Users Guide. NISTIR 6784, 2002 Ed. U.S. Department of Commerce. In: National Institute of Standards and Technology, Gaithersburg, MD

70. Ng KT, Gardner L (2007) Buckling of stainless steel columns and beams in fire. Eng Struct 29(5):717-730

71. Ngu CN (2004) Calcination of Gypsum Plasterboard under Fire Exposure. Fire Engineering Research Report 04/6. Department of Civil Engineering, University of Canterbury, Christchurch, New Zealand.

72. Outinen J (2006) Mechanical Properties of Structural Steel at High Temperatures and after Cooling Down, Doctoral Dissertation. Helsinki University of Technology, Espoo, Finland.

73. Outinen J, Mäkeläinen P (1999) Mechanical Properties of an Austenitic Stainless Steel at Elevated Temperatures. In: Chan SL, Teng JG (eds) Advances in Steel Structures (ICASS `99), Elsevier, Oxford. pp 1063-1069 doi:10.1016/b978-008043015-7/50124-x

74. Outinen J, Mäkeläinen P (2004) Mechanical properties of structural steel at elevated temperatures and after cooling down. Fire Mater 28(2¿4):237-251

75. Quintiere JG (1989) Fundamentals of Enclosure Fire ¿Zone¿ Models. J Fire Protect Eng 1(3):99-119

76. Rahmanian I (2011) Thermal And Mechanical Properties of Gypsum Boards And Their Influence on Fire Resistance Of Gypsum Board Based Systems. University of Manchester, Manchester, England.

77. Ramberg W, Osgood WR (1943) Description of Stress¿Strain Curves By Three Parameters. Technical Notes. National Advisory Committee for Aeronautics, Washington, DC.

78. Ranawaka T, Mahendran M (2006) Finite element analyses of cold-formed steel columns subject to distortional buckling under simulated fire conditions, Proceedings of the International Colloquium on Stability and Ductility of Steel Structures. Instituto Superior Técnico, Lisbon, Portugal.

79. Ranawaka T, Mahendran M (2009) Experimental study of the mechanical properties of light gauge cold-formed steels at elevated temperatures. Fire Safety J 44(2):219-229 doi:10.1016/j.firesaf.2008.06.006

80. Ranawaka T, Mahendran M (2009) Distortional buckling tests of cold-formed steel compression members at elevated temperatures. J Construct Steel Res 65(2):249-259 doi:10.1016/j.jcsr.2008.09.002

81. Ranawaka T, Mahendran M (2010) Numerical modelling of light gauge cold-formed steel compression members subjected to distortional buckling at elevated temperatures. Thin-Walled Structures 48(4¿5):334-344 doi:10.1016/j.tws.2009.11.004

82. Rasmussen KJR (2003) Full-range stress¿strain curves for stainless steel alloys. J Construct Steel Res 59(1):47-61 [http://dx.doi.org/10.1016/S0143-974X(02)00018-4]   http://dx.doi.org/10.1016/S0143-974X(02)00018-4

83. Rubini P (2006) Simulation of Fires in Enclosure (SOFIE). Cranfield University, Bedfordshire, United Kingdom.

84. Santos P, Martins C, da Silva LS, Bragança L (2013) Thermal performance of lightweight steel framed wall: The importance of flanking thermal losses. J Build Phys doi:10.1177/1744259113499212

85. Schafer BW (2006) Review: The Direct Strength Method of Cold-Formed Steel Member Design. Proceedings of the Stability and Ductility of Steel Structures. Instituto Superior Técnico, Lisbon, Portugal.

86. Schafer BW (2011) Cold-formed steel structures around the world: A review of recent advances in applications, analysis and design. ECCS, Steel Construct 4(3):141-149 doi:10.1002/stco.201110019

87. Schafer BW, Peköz T (1998) Computational modeling of cold-formed steel: characterizing geometric imperfections and residual stresses. J Construct Steel Res 47(3):193-210

88. Schafer BW, Li Z, Moen CD (2010) Computational modeling of cold-formed steel. Thin-Walled Structures 48(10¿11):752-762 [http://dx.doi.org/10.1016/S0143-974X(02)00018-4] http://dx.doi.org/10.1016/S0143-974X(02)00018-4

89. (1993) Building Design using Cold Formed Steel Sections: Fire Protection. British Library Cataloguing-in-Publication Data, Silwood Park, Ascot.

90. Shahbazian A, Wang YC (2011) Application of the Direct Strength Method to local buckling resistance of thin-walled steel members with non-uniform elevated temperatures under axial compression. Thin-Walled Structures 49(12):1573-1583

91. Shahbazian A, Wang YC (2011) Calculating the global buckling resistance of thin-walled steel members with uniform and non-uniform elevated temperatures under axial compression. Thin-Walled Structures 49(11):1415-1428 doi:10.1016/j.tws.2011.07.001

92. Shahbazian A, Wang YC (2012) Direct Strength Method for calculating distortional buckling capacity of cold-formed thin-walled steel columns with uniform and non-uniform elevated temperatures. Thin-Walled Structures 53:188-199

93. Shahbazian A, Wang YC (2013) A simplified approach for calculating temperatures in axially loaded cold-formed thin-walled steel studs in wall panel assemblies exposed to fire from one side. Thin-Walled Structures 64:60-72

94. Spalding B (1978) PHOENICS. Concentration Heat And Momentum Limited, London,UK.

95. (1998) AS 4100-1998: Steel Structures. Standards Australia Limited, Sydney, Australia.

96. Sultan M (1996) A model for predicting heat transfer through noninsulated unloaded steel-stud gypsum board wall assemblies

exposed to fire. Fire Technol 32(3):239-259 doi:10.1007/bf01040217

97. Sultan MA, Lougheed GD (2002) Results of Fire Resistance Tests on Full-Scale Gypsum Board Wall Assemblies. Internal Report, Institute for Research in Construction, vol 833. National Research Council Canada, Ottawa, ON.

98. Sultan MA, Séguin YP, Leroux P (1998) Results of fire resistance tests on full-scale floor assemblies, IRC-IR-764. National Research Council Canada, Ottawa, ON.

99. Thomas G (2002) Thermal properties of gypsum plasterboard at high temperatures. Fire Mater 26:37-45 doi:10.1002/fam.786

100. (2003) UL 263: Standard for Safety for Fire Tests of Building Construction and Materials. Underwriters Laboratories Inc, Northbrook, IL.

101. Uribe J (1969) Aspects of The Effects of Cold-Forming on The Properties And Performance of Light-Gage Structural Members. Department of Structural Engineering, Cornell University, Ithaca, NY.

102. Wakili KG, Hugi E (2009) Four types of gypsum plaster boards and their thermophysical properties under fire condition. J Fire Sci 27(1):27-43 doi:10.1177/0734904108094514

103. Wang YC, Davies JM (2000) Design of thin-walled steel channel columns in fire using Eurocode 3 Part 1.3. Proceedings of the 1st International Workshop on Structures in Fire, Copenhagen, Denmark.

104. Yan S, Young B (2011) Tests of single shear bolted connections of thin sheet steels at elevated temperatures¿Part I: Steady state tests. Thin-Walled Structures 49(10):1320-1333

105. Yan S, Young B (2011) Tests of single shear bolted connections of thin sheet steels at elevated temperatures¿Part II: Transient state tests. Thin-Walled Structures 49(10):1334-1340

106. Yan S, Young B (2012) Bearing factors for single shear bolted connections of thin sheet steels at elevated temperatures. Thin-Walled Structures 52:126-142

107. Yan S, Young B (2012) Screwed connections of thin sheet steels at elevated temperatures ¿ Part I: Steady state tests. Engineering Structures 35:234-243

108. Yan S, Young B (2012) Screwed connections of thin sheet steels at elevated temperatures ¿ Part II: Transient state tests. Engineering Structures 35:228-233

109. Zeinoddini VM, Schafer BW (2012) Simulation of geometric imperfections in cold-formed steel members using spectral representation approach. Thin-Walled Structures 60:105-117

Chapter 8

Organic-Inorganic Materials Containing Nanoparticles of Zirconium Hydrophosphate for Baromembrane Separation

Yuliya S Dzyazko[1], Ludmila M Rozhdestvenskaya[1], Yu G Zmievskii[2], Alexander I Vilenskii[3], Valerii G Myronchuk[2], Ludmila V Kornienko[2], Sergey V Vasilyuk[1], and Nikolay N Tsyba[4]

[1]Department of Sorption and Membrane Materials and Processes, V.I. Vernadskii Institute of General and Inorganic Chemistry, NASU, Palladin Pr. 32/34, Kiev, 03142, Ukraine

[2]Department of Process Equipment and Computer Technology Design, National University of Food Technologies of the Ministry of Education and Science of Ukraine, Vladimirskaya str. 48, Kiev, 01601, Ukraine

[3]Department of Membrane Technologies, A.V. Shubnikov Institute of Crystallography, RAS, Leninskii pr. 59, Moscow 119333, Russian Federation

[4]Department of Carbon Sorbents for Medical and Ecological Application, Institute for Sorption and Problems of Endoecology, NASU, General Naumov Str. 13, Kiev, 03163, Ukraine

ABSTRACT

Organic-inorganic membranes were obtained by stepwise modification of poly(ethyleneterephthalate) track membrane with nanoparticles of zirconium hydrophosphate. The modifier was inserted inside pores of the polymer, a size of which is 0.33 μm. Inner active layer was formed by this manner. Evolution of morphology and functional properties of the membranes were investigated using methods of porosimetry, potentiometry and electron microscopy. The nanoparticles (4 to 10 nm) were found to form aggregates, which block pores of the polymer. Pores between the aggregates (4 to 8 nm) as well as considerable surface charge density provide significant transport numbers of counter ions (up to 0.86 for Na^+). The materials were applied to baromembrane separation of corn distillery. It was found that precipitate is formed mainly inside the pores of the pristine membrane. In the case of the organic-inorganic material, the deposition occurs onto the outer surface and can be removed by mechanical way. Location of the active layer inside membranes protects it against damage.

BACKGROUND

Application of ultrafiltration involves a wide variety of fields, for instance, recovery of ionic species (usually enhanced by polyelectrolytes) [1],[2], treatment of brackish [3] and waste water [4], food industry (for juice concentration [5], protein recovery from whey [6]) and medicine [7]. Both polymer and ceramic membranes are used for baromembrane processes [8].

Almost all the commercially available membranes contain thin nanoporous active layer applied to macroporous substrate. A thickness of the active layer is up to several micrometers. The active layer is necessary to provide separation ability of the membranes, and the macroporous substrate guarantees their low hydrodynamic resistance. In the case of polymer membranes, the active layer is formed, particularly by interfacial polycondensation, plasma polymerization, in situ polymerization at the outer surface of the membrane, polymer grafting [9]. Sol-gel method is often applied to the formation of active layer of inorganic membranes [10].

In opposite to fragile inorganic materials, polymer separators are more attractive for operation due to their elasticity and stability of small pores, which determine permittivity of the membranes, against high pressure. However, foiling of the membranes by organic species as well as development of microorganism debris inside the polymer decreases a lifetime of the membranes on the one hand and declines permeate fluxes on the other hand [11],[12]. In order to minimize fouling with organics and microorganisms, insertion of nanoparticles of inorganic compounds, such as SiO_2[13],[14], particularly stabilized with N-halamine [14], ZrO_2[15], Fe_2O_3 stabilized with chitosan [16] and TiO_2[17], into polymers has been proposed. Two approaches were used for the preparation of ultrafiltration [13]-[16] and reverse osmotic membranes: insertion of sol or suspension containing the inorganic constituent into the dissolved polymer or vice versa [13]-[16] as well as modification of the polymer membrane, which had been prepared preliminary [17]. These approaches require further coupling of the obtained film with macroporous substrate or use of polymer composite membranes consisting of the substrate and active layer. Another problem is a purposeful formation of needed porosity. In the case of modification of preliminary formed polymer membrane, a question of necessity of multiple modification is still opened.

Moreover, fouling of the membranes requires their periodical cleaning, which is often carried out mechanically or by means of hydrodynamic pulsation [11],[12]. This causes damages of thin active layer and, as a result, shortage of their lifetime. At last, complex and expensive equipment is needed for industrial manufacture of the composite membranes.

Earlier electrodialysis membranes were obtained by formation of active layer inside macroporose ceramics. ZrO_2 nanoparticles were found to block macropores of the membrane and form secondary porosity [18]. Pores between these particles as well as high surface charge density provide semipermittivity of the membranes towards anions in acidic media and towards cations in alkaline solutions [19],[20]. This gives a possibility to assume a similar approach to create polymer-based organic-inorganic membranes also for ultrafiltration.

In this work, the membranes were tested by deionized water and corn distillery. In the last case, the ultrafiltration allows us to remove useful components (crude proteins, fat etc.) [21], which can be further used for preparation of livestock feed. Simultaneously, ecological problem of wastewater purification can be solved.

EXPERIMENTAL

Track Membranes

Track membrane has been chosen as a model polymer matrix since its porous structure involves through regular pores, a size of which is several hundreds nanometers [22]. Studies were performed using a poly (ethyleneterephthalate) (PETP), a thickness of which was 11 μm. Preliminary, the film was irradiated with Xe ions with an energy of 1 MeV/nucleon and a density of 2×10^9 ions cm^{-2} under the vacuum environment of 10^{-6} Torr similar to [23], [24]. The energy of incident ions was sufficient to form through latent tracks. Then, UV sensibilisation was carried out for restructurisation of fragments of molecular compounds in order to shorten the period of subsequent etching. The etching was performed in a KOH solution (250 mol m^{-3}) at 348 K.

Modification of the Polymer Matrix

Polymer matrix was filled with zirconium hydrophosphate (ZHP), a choice of the modifier due to its chemical stability and possibility to obtain nanosized particles inside polymer pores [25]-[28]. In opposite to hydrated zirconium dioxide, which was used for the modification of ceramics, ZHP is characterized by higher surface charge density in neutral solutions. In prospect, the membranes can be used for other tasks, which require this property. Moreover, a treatment of the immersed polymer with an alkaline solution for deposition of hydrated zirconium dioxide can result in damage of the membrane material.

Sol of insoluble zirconium hydroxocomplexes was prepared and analysed as described earlier [18]. The membrane was boiled in deionized water under vacuum, treated with a H_3PO_4 solution (1,000 mol m^{-3}), dried at ≈ 298 K and heated at 343 K, the ion-exchanger was removed from the outer surface of the membrane by means of ultrasonic activation at 30 kHz using a Bandelin device (Bandelin

Electronic GmbH & Co. KG, Berlin, Germany). Then, the membranes were dried at 343 K down to constant mass, weighted and stored in a desiccator over $CaCl_2$. The modification was repeated several times, after each modification cycle a sample was taken for investigations.

Morphology and Porosity of the Membranes

Both outer surface and transverse section of the membranes were investigated using a JEOL JSM-6060 LV scanning electron microscope (JEOL Ltd., Akishima-shi, Japan), elementary analysis of the modifier incorporated into the polymer was provided by this manner. Preliminarily, the samples were coated with an ultrathin gold layer at 3 Pa by means of an auto fine coater JEOL JFC-1600 (JEOL Ltd.).

A fine-dispersed powder was obtained from the composite by its grinding under cooling with liquid nitrogen. The powder was researched using a JEOL JEM 1230 transmission electron microscope (JEOL Ltd.).

Micro- and mesopores were determined by means of nitrogen desorption using a Quantachrome Autosorb 6B analyzer (Quantachrome instruments, Boynton Beach, FL, USA). Before the measurements, the samples were vacuumized at 343 K. Bulk density ρb was estimated from mass and geometrical sizes of the membrane, particle density was found with a picnometer (Archimedes) method similar to [29].

Total porosity (ε) was calculated as $1-\dfrac{pb}{pb}1-\dfrac{pb}{pb}$.

Ion-Exchange Capacity and Membrane Potential

Cation-exchange capacity of the membranes was determined by their multiple treatment with a NaCl solution (100 mol m^{-3}), washing with deionized water (electrical conductivity of the effluent was performed), treatment with a HCl solution (100 mol m^{-3}) and analysis of the effluent using a PFM-U flame photometer.

Membrane potential was measured at 298 K using a two-compartment divided cell similar to [30],[31]. Pairs of NaCl solutions (0.05 to 5 and 10 mol m^{-3}) filled their chambers, where Ag/AgCl electrodes were placed.

Separation Process

Experimental set-up involved a plane membrane cell, liquid line, thermostat and pressure and flow controllers (Figure 1). The effective membrane area was 2.1×10^{-3} m².

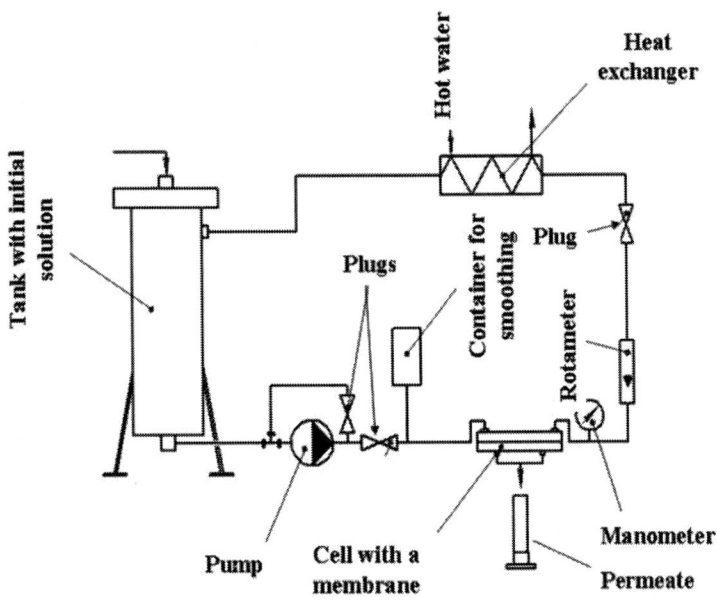

Figure 1: Experimental set-up for baromembrane processes.

Preliminary testing was as follows. Initially deionized water was passed through the membrane at 0.3 MPa and 333 K for 16 h. After the crimping by this manner, the membrane was stored at room temperature and atmosphere pressure for 24 h. Then, the passage was continued in order to determine the resistance of the membrane. After this, water was replaced by corn distillery, which had been preliminary centrifuged and filtered using Buchner funnel. The separation was performed for 4 h (the pressure was kept at 0.1 or 0.3 MPa), then the liquid was replaced by deionized water to find the membrane resistance again. After this, the membrane was removed from the cell, cleaned, dried and investigated with SEM and porosimetry methods. A content of the matters in the permeate and concentrate was determined with a refractory method.

The membranes (both the pristine one and just after modification) were also tested several times. First of all, the crimping was performed as described above. The separation cycle was as follows. Corn distillery was passed through the system at 0.3 MPa for 4 h. When the separation process was finished, the membrane was removed from the cell, its outer surface was cleaned mechanically and washed with deionized water. The membrane was stored in aqueous medium for 20 h. After this, the membrane was inserted into the cell again and tested with deionized water. The separation cycles were repeated five times. Then, the membrane was stored in deionized water for 96 h, washed with a 0.1 M HCl solution and water up to neutral reaction of the effluent. The separation cycle was carried out again.

RESULTS AND DISCUSSION

Morphology and Porosity of the Pristine Membranes

SEM images of surface and cross-section of the pristine track membrane are represented in Figure 2. Round holes of regular shape are seen at the surface, the size of the holes is 0.33 μm. In general, no roughness is visible around the circumference of the holes indicating evidently smooth walls of the pores. A distance between holes is up to several micrometers. Some holes are double and even triple. Though pores are seen in the SEM image of a cross-section, some pores show tortuosity (however, most of them are straight), some of them merge and branch. Assuming cylindrical and regular shape of the pores, the porosity has been estimated as 0.1 by means of analysis of ten images. This in an agreement with data obtained with a picnometer method (Table 1).

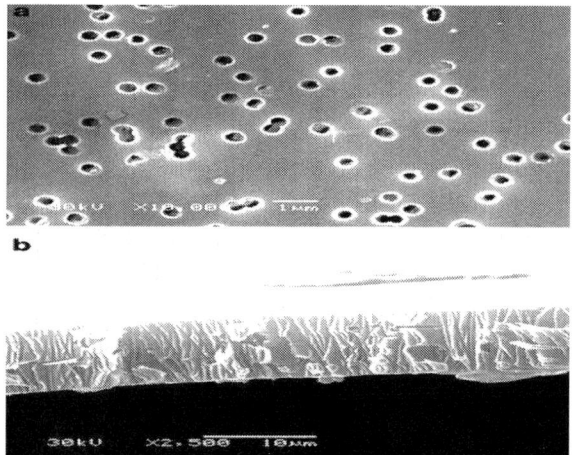

Figure 2: SEM image of outer surface (a) and cross-section (b) of the pristine membrane. Through pores, a shape of which can be assumed as cylindric, are visible.

Table 1: Characteristics of the membranes

m	ε	Volume of micropores, cm^3g^{-1}	S, m^2g^{-1}	A × 10^3, mmol g^{-1}	η, C m^{-2}	r, nm
0	0.109	7.01×10^{-5}	1.9	5.9	0.030	158
0.047	0.082	2.32×10^{-4}	8.2	1.5	0.018	4.8
0.052	0.080	3.35×10^{-4}	11.8	2.3	0.019	3.1
0.056	0.075	3.86×10^{-4}	13.6	2.8	0.020	2.2
0.061	0.070	4.57×10^{-4}	16.1	3.3	0.020	3.2
0.063	0.066	5.06×10^{-4}	17.9	3.8	0.021	2.7

Dzyazko et al.

Dzyazko et al. Nanoscale Research Letters 2015 10:64, doi:10.1186/s11671-015-0758-x

Differential pore size distribution is given in Figure 3. Two peaks are visible: the first one as well as micropores corresponds to pore radius (r) up to 4.5 nm and evidently related to polymer heterogeneities. The second peak is attributed through pores, they are partially outside the region of sensitivity of the method. Wide peak is evidently due to tortuosity and merger-branching of the pores.

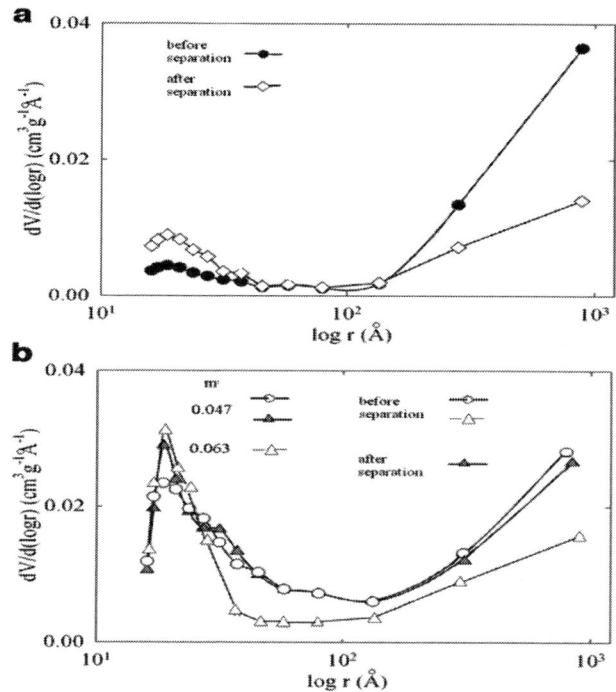

Figure 3: Differential pore size distributions. Obtained for pristine polymer (a) and organic-inorganic membrane containing 4.7 and 6.3 mass % ZHP (b). The pristine (a) and composite (4.7% ZHP) (b) membranes were investigated before and after the separation process.

Morphology and Porosity of the Modified Membranes

As shown earlier with methods of dynamic laser light scattering and TEM, sol of insoluble zirconium hydroxocomplexes includes both single globular nanoparticles, a minimal size of which is 4 nm, and their aggregates [18]. The particles with a diameter of 15 (non-aggregated globules) and 120 nm dominate in sol. Pores of the pristine membrane are available both for nanoparticles and their aggregates.

Stepwise modification, which involves removal of the precipitate from outer surface of the membranes, results in an increase of ZHP content inside the polymer (see Table 1). The largest growth of mass fraction (m) of the inorganic constituent is reached during the first modification cycle. Smaller increase of the m value is achieved during further modification, no sufficient growth of ZHP amount has been found after the fifth cycle. In owing to this, no further modification was performed.

A major part of holes becomes invisible in SEM image of the outer surface of the modified sample (Figure 4). Some round convexities are seen. Larger size and larger distance between them than for the pristine membrane indicate blocking and stretching of the macropores and their partial squeezing from the side of filled pores.

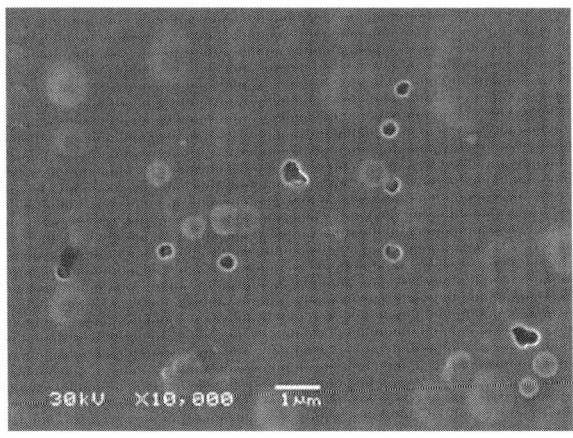

Figure 4: CEM image of the organic-inorganic membrane. Pores attributed to the polymer are visible only partially.

As follows from Table 1, increasing of ZHP amount in the polymer results in a growth of microporosity and specific surface area (S), the total porosity decreases simultaneously. Differential pore size distribution shows a higher peak at $r = 1.7$ nm in a comparison with that of the pristine membrane. Moreover, the second narrow peak at $r = 3$ nm is visible for the membrane with a minimal ZHP content. This peak indicates a presence of larger particle than those which form smaller pores. Indeed, TEM image of the membrane powder shows the agglomerate, which consists of aggregates, a size of which is from 30 nm (Figure 5). The aggregates include smaller nanoparticles. The peak at $r = 3$ nm practically disappears for the membrane with a maximal content of the modifier.

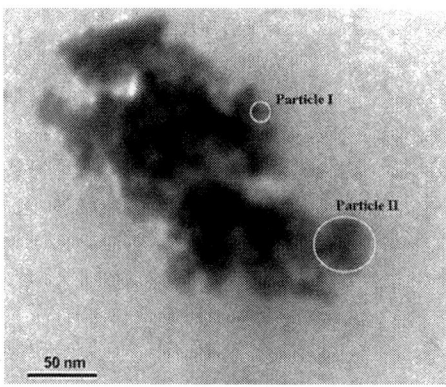

Figure 5: TEM image of ZHP agglomerate incorporated into the polymer. The nanoparticles, which form aggregates, are visible.

Regarding to the pristine membrane, its cation-exchange capacity (A, see Table 1) is caused by $-COOH$ groups, which are formed during etching of the polymer by alkaline solution [23]. Insertion of ZHP into the polymer predictably causes increases of capacity. It should be noted, that a Zr:P molar ratio was $\approx 1:1.9$ for all the samples, this is rather close to that for crystalline material (α-ZHP modification [32]). Moreover, the membranes demonstrate an increase of exchange capacity with increasing of the modifier amount.

Incorporated Modifier

Microporosity of the membranes is undoubtedly attributed to the modifier. In order to estimate loosening-compactness of the porous structure of the modifier on the level of micropores, the α and β parameters have been proposed. The α parameter is a $\frac{m_n}{m_1}$ ratio, where the '1' index corresponds to the one-time modified membrane (i.e. to minimal ZHP content), 'n' is related to membranes containing larger ZHP amount. Similarly, the β parameter corresponds to a $\frac{V_{micr,n}}{V_{micr,1}}$ ratio, where V_{micr} is a volume of micropores. Regarding the membrane with a minimal content of the modifier, $\alpha = \beta = 1$. In our case, the $\beta - \alpha$ plot is linear (Figure 6). Since $\frac{d_\beta}{d_\alpha} > 1$ ($\frac{d_\beta}{d_\alpha} = 3.3$), stepwise modification causes loosening of porous structure of the filler due to deposition of more friable microporous formations from cycle to cycle.

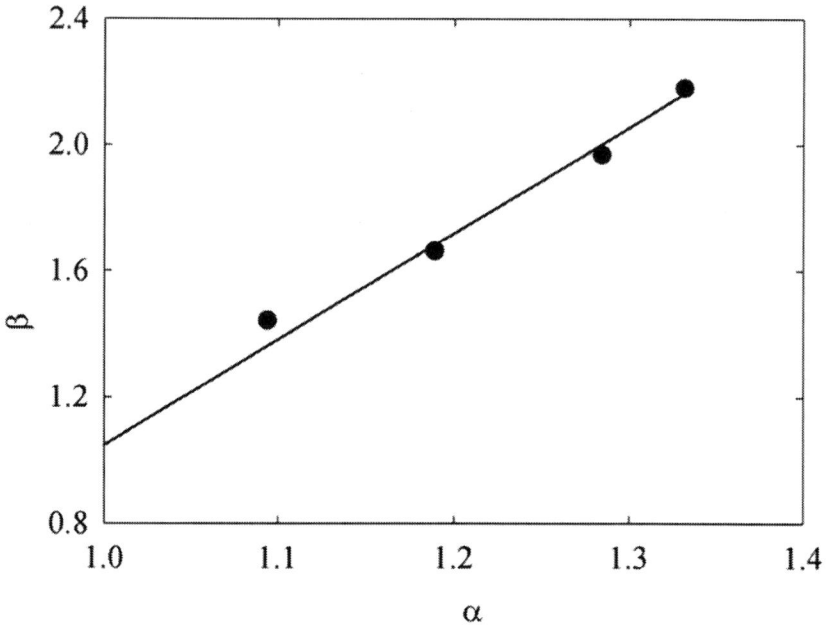

Figure 6: β parameter as a function of α parameter.

In a framework of the first approximation, bulk density of the incorporated modifier (ρ'_b) can be determined from the m value and

decrease of porosity. The porosity of ZHP (ε') was calculated as $1 - \frac{\rho'_b}{\rho'_p}$, where ρ'_p is the particle density (3.3 g cm^{-3} for crystalline α-ZHP modification [32]). The plots of $\rho'_p - m$ and $\varepsilon' - m$ demonstrate the maximum and minimum, respectively (Figure 7), which is evidently a result of a contradiction of two reasons: increase of microporosity on the one hand and decrease of mesopore volume on the other hand.

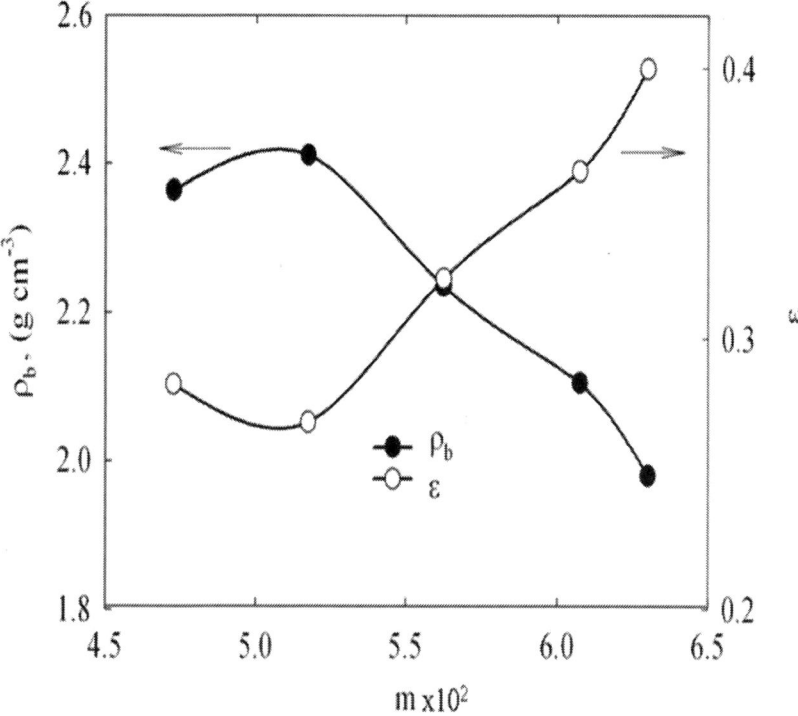

Figure 7: Bulk density of the inner ZHP layer and its porosity. These characteristics are given as functions of mass fraction of the modifier.

Specific surface area of incorporated ZHP demonstrates a growth with increasing of the modifier content evidently due to development of microporosity (Figure 8). Diameter of the globules was calculated as

$\bar{d} = \dfrac{6}{\rho_i S_i}$. As seen from the figure, effective diameter of the particles decreases with increase of the m value indicating deposition of the smallest particles inside the membranes from stage to stage of the modification.

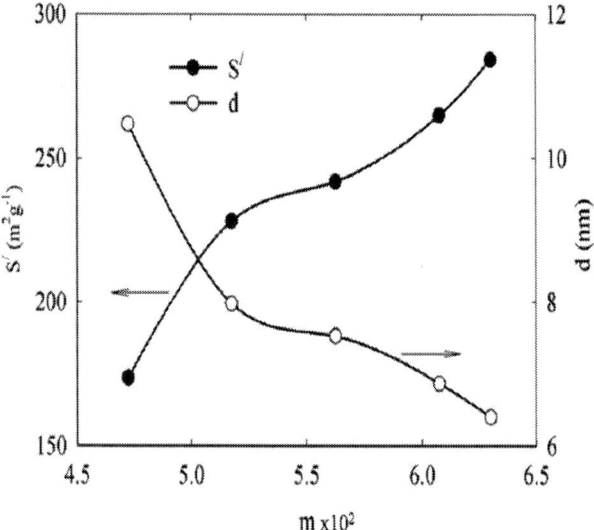

Figure 8: Specific surface area of the modifier and particle diameter. These characteristics are given as functions of mass fraction of the modifier.

Potentiometric Transport Numbers

In all the cases, the membrane potential was registered. Regarding the pristine membrane, it contains – COOH groups, which are formed during etching of the polymer with alkaline solution. These groups are dissociated partially in neutral media. An excess of counter ions (Na^+) in the diffusion parts of electric double layer causes its slightly expressed charge-selective properties towards cations. In the case of organic-inorganic membranes, the membrane potential is due to the dissociation of $(-O)_2PO_2H$ and $-OPO_3H_2$ groups:

$$(-O)_2PO_2H \rightarrow (-O)_2PO_2^- + H^+ \tag{1}$$

$$-OPO_3H_2 \rightarrow -OPO_3H^- + H^+ \tag{2}$$

$$-OPO_3H^- \rightarrow -OPO_3^{2-} + H^+ \tag{3}$$

In the last case, the transport number of counter-ions ($\bar{t}$) through the membrane was determined from the data of membrane potential (E_m) according to the formula for 1,1 binary electrolyte [31]:

$$E_m = \frac{RT}{F}\left[\ln\frac{a_2}{a_1} \pm 2\int_{a_1}^{a_2}(1-\bar{t})d\ln a_{\pm}\right]$$

(4)

where a_1 and a_2 are the activities of counter-ions in less and more concentrated solutions, respectively, $a_{\pm}$ is the activity of the solution of varied concentration (more concentrated solution in our case), R is the gas constant, F is the Faraday constant and T is the temperature. The transport numbers of Na^+ ions are represented in Figure 9, they are sensitive to the solution concentration and approximated to the 'true' value with a decrease of a difference of the solution concentration [30]. This value is evidently realized under applied potential.

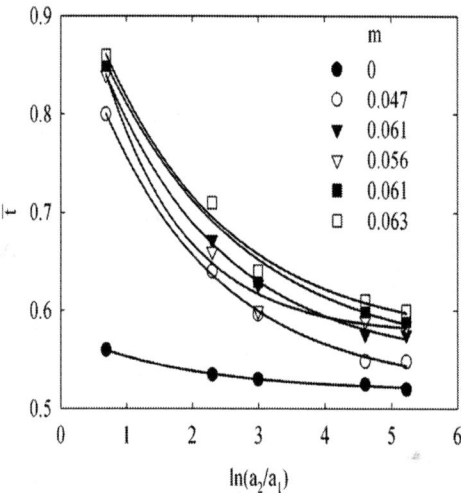

Figure 9: Transport number of counter-ions as a function of $n\frac{a_1}{a_2}$.

A radius of pores, which provide the membrane potential, can be calculated from the expression [34]:

$$\bar{t} = t\left(1 + \frac{F\bar{r}C}{k\bar{\eta}}\right)\left(t + \frac{F\bar{r}C}{k\bar{\eta}}\right)^{-1}$$

(5)

where t is the transport number of counter-ions (Na^+) in a solution (0.4), k is the shape coefficient (k = 2.8 for pores formed with globules), η is the surface charge density and C is the solution concentration. The surface charge density (see Table 1) was found as $\frac{FA}{S}$.

Equation 5 gives the transport number, at which the concentrations of the solutions from two sides of the membranes are close to each other. In other words, extrapolation of the curve like $r - \ln\frac{a_1}{a_2}$ to the ordinate axis allows us to obtain 'true' r magnitude (Figure 10). These data are shown in Table 1. The result obtained for the pristine membrane is in a good agreement with SEM observation. However, in the case of modified membranes, the potentiometric method gives nanosized values, which are in a contradiction with porosimetric measurements (they show a presence of larger pores).

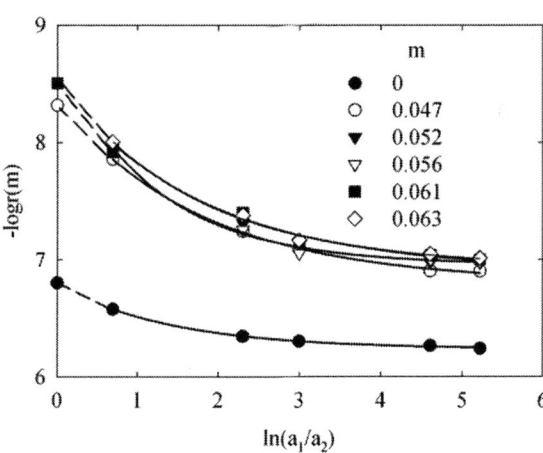

Figure 10: Logarithm of pore radius as a function of $\ln\frac{a_1}{a_2}$. Calculations were performed according to Equation 5.

Table 2: Hydrodynamic resistance of the membrane and their selectivity (preliminary testing)

	Pretreatment with water, R, m⁻¹		Separation, φ, %		Treatment with water after separation, R, m⁻¹	
m	$\tau = 0$	$\tau = 16h$	$\Delta P = 0.1MPa$	$\Delta P = 0.3MPa$	$\Delta P = 0.1MPa$	$\Delta P = 0.3MPa$
0	0.11×10^{13}	1.81×10^{13}	31	31	4.35×10^{13}	5.25×10^{13}
0.047	0.81×10^{13}	2.13×10^{13}	31	34	4.85×10^{13}	5.75×10^{13}

Dzyazko et al.

Dzyazko et al.Nanoscale Research Letters 2015 10:64, doi:10.1186/s11671-015-0758-x

Thus, a mechanism of filling of the polymer matrix with pores, which are smaller than 1 μm, is similar to those for ceramics ($r > 1$ μm). Matrix pores are blocked with aggregates of ZHP nanoparticles during the first synthesis stage (Figure 11). The aggregates evidently give pores, a radius of which is about 4 nm (see Figure 4). The aggregates isolate wide cavities, which are partially seen in the differential pore size distributions. During further modification stages, only nanosized particles of sol are able to penetrate inside matrix pores. Pores between the aggregates are gradually blocked with ZHP nanoparticles, making full filling of macropores of the polymer impossible. Since the modifier occupies about 30% of the total pore volume, its maximal thickness is ≈ 3 μm (assuming that all the modifier form 'corks').

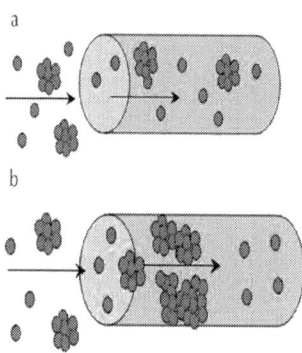

Figure 11: Filling of the membrane pores during the first (a) and further (b) stages of modification. The nanoparticles inside the polymer macropores block them and form secondary porosity.

Baromembrane Separation

In order to characterize the membrane behaviour during the process, the experimental data were analyzed as follows. A flux (J) of permeate was determined as [8]:

$$J = \frac{3,600V}{A\tau}$$

(6)

Here, V is the permeate volume, A is the effective membrane area and τ is the time. Selectivity of the membranes (φ) was estimated according to the expression:

$$\varphi = \left(1 - \frac{C_2}{C_1}\right) \times 100\%$$

(7)

where C_1 and C_2 are the concentration of species in concentrate and permeate, respectively. At last, hydrodynamic resistance (R) of the membrane was calculated according to Darcy equation:

$$J = \frac{\Delta P}{\mu R}$$

(8)

where μ is the dynamic viscosity and ΔP is the pressure drop.

Hydrodynamic resistance towards water is predictably higher for the membrane containing ZHP (m = 0.047) than that for the pristine membrane (Table 2). Other materials showed the resistance, which was higher in two times in a comparison with the pristine separator, these materials were not used for testing. As seen from the table, a flux of water through the membrane tends to increase with a growth of pressure drop. In a contrary, a flux of the permeate obtained during separation of corn distillery decreases with increasing of pressure (Figure 12). Moreover, the J value decreases in time due to fouling. The most stable flux has been found for the modified membrane at 0.3 MPa. In other cases, the flux gradually decreased in time. No sufficient difference of selectivity was found for the modified membrane in a comparison with the pristine separator (see Table 2). It means less size of species in the distillery (<8 nm) than pore size of the membrane. These species are able to penetrate through the membrane into permeate.

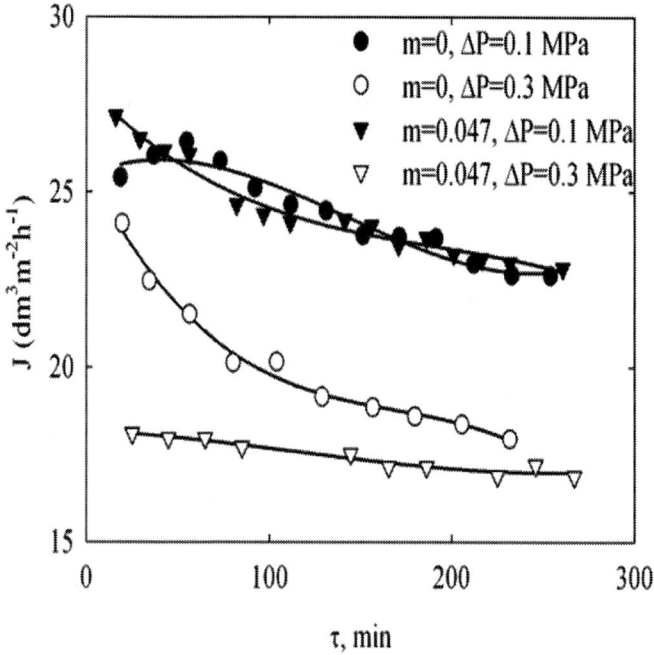

Figure 12: Flux of permeate through pristine and modified membranes as a function of time.

After the end of the processes, higher amount of precipitate was found on the outer surface of the modified membrane, the membrane shows higher hydrodynamic resistance. No sufficient difference of pore size distribution was found for the modified membrane (see Figure 3). In opposite to composite material, the pristine membrane demonstrates considerable increase of porosity due to pores with $r < 4.5$ nm. This growth is evidently caused by particles of organics, which are deposited inside pores. At the same time, a volume of through pores is considerably lower in comparison with that for the membrane, which is free from a precipitate. Moreover, a change of morphology of the polymer membrane is seen in the SEM image (Figure 13, compare with Figure 2). On the other hand, the images of the modified membrane before and after the process are practically the same.

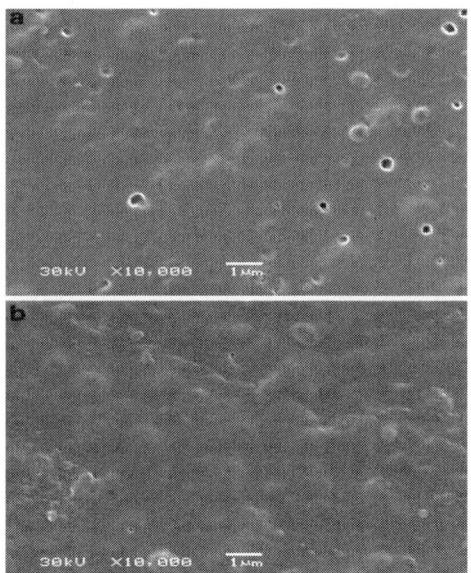

Figure 13: Morphology of the modified (a) and pristine (b) membranes after separation.

Thus, in the case of the pristine membrane, a decrease of the flux is evidently caused by the precipitate formation inside the membrane. Regarding the modified separator, the fouling is evidently due to deposition onto outer surface. This precipitate can be easy removed by mechanical way or by hydrodynamic pulsation, thus, the organic-inorganic membrane looks more attractive from the practical point of view. Location of the active layer inside the membrane prevents its damage.

The data obtained for repeated testing are given in Table 3. In the case of pristine membrane, hydrodynamic pressure increases dramatically just after the first operation cycle. Cleaning of the membrane did not provide removal of the precipitate from pores. No considerable growth of this parameter was found for the modified membrane during the next operation cycles. Further change of the resistance is within the statistical error indicating fouling mainly during the first cycle. The modified membrane demonstrates lower resistance. If no removal of the precipitate from outer surface of the separator was provided, a ratio of resistances of the modified and pristine membrane is 1.2 (see Table 2). As follows from Table 3, this ratio becomes 0.5 after cleaning. It means the modified membrane accumulates organics only onto outer surface. Regarding the pristine membrane, both outer surface and pores are poisoned.

Table 3: Hydrodynamic resistance of the membrane and their selectivity (repeated testing at 0.3 MPa)

Cycle number	Pristine membrane		Modified membrane (m = 0.047)	
	Selectivity,φ, %	Treatment with water after separation and cleaning,R, m^{-1}	Selectivity,φ, %	Treatment with water after separation and cleaning,R, m^{-1}
1	31	5.05×10^{13}	34	2.53×10^{13}
2	32	5.15×10^{13}	35	2.66×10^{13}
3	32	5.18×10^{13}	36	2.68×10^{13}
4	32	5.21×10^{13}	36	2.69×10^{13}
5	32	5.23×10^{13}	36	2.70×10^{13}
Chemical regeneration				
6	32	5.07×10^{13}	34	2.56×10^{13}

Dzyazko et al.

Dzyazko et al.Nanoscale Research Letters 2015 10:64, doi:10.1186/s11671-015-0758-x

However, after the fifth cycle followed by long-time storage in deionized water, colonies of microorganisms were found on the outer surface of the pristine and modified membranes. This is evidently due to adhesion of microorganisms during separation process. Adhesion is possible on the outer surface, since their penetration inside membranes is difficult due to steric factor. In owing to this, the membranes were treated with a HCl solution (see subsection 'Separation process'). After cleaning, both selectivity of the membranes and their hydrodynamic resistance have been found to be close to those for the first cycle due to the removal of the precipitate from pores.

Thus, the advantage of the modified membrane is its lower resistance evidently due to stability against accumulation of organics inside pores. However, long-time operation requires also protection of the outer surface of the membranes or their regular disinfection. A solution of the problems is outside the scope of this work.

CONCLUSIONS

Modification of polymer track membranes, which was performed by insertion of inorganic filler like ZHP inside their macropores, allows us to obtain the inner active layer in opposite to majority of commercially available membranes. The mechanism of stepwise modification is as follows. First, the macropores of the polymer are blocked with aggregates of nanoparticles a size of which is 10 nm. The 'corks' isolate wide cavities and provide permittivity of the membrane towards cations, as shown by measurements of membrane potential. No considerable increase of the modifier amount was found after further modification stages, since secondary porosity limits ZHP deposition inside the polymer.

Both the pristine and composite membranes were tested for baromembrane separation of corn distillery. In the case of modified separator, precipitation occurs directly onto the outer surface in opposite to the pristine membrane, for which deposition inside pores was found. The precipitate can be easily removed from the surface. Location of the active layer inside membrane prevents its mechanical damage.

The directions of further investigations are evidently purposeful regulation of the filler amount inside the membranes, establishment of interrelation between this characteristic and functional properties of the membranes, modification of different types of porous polymers and application of the composites to solution of different tasks of baromembrane separation. Moreover, the protection of outer surface of the membranes against biogenic fouling or regular disinfection is needed to provide their long lifetime, especially in media of liquids of biological origin.

AUTHORS' CONTRIBUTION

A task of the work was formulated by YSD and YGZ. YSD occupied investigations with TEM and SEM methods and analysis, and summarizing of the experimental data obtained during researches of structure and functional properties of the membranes is done by LMR and SLV. AIV was responsible for preparation of track membranes.

Porosimetric measurements were carried out by NNT. Investigations of baromembrane separation were performed by YGZ, VGM and LVK. All authors read and approved the final manuscript.

ACKNOWLEDGEMENTS

The work was supported by projects within the framework of programs supported by the government of Ukraine 'Nanotechnologies and nanomaterials' (grant no. 6.22.1.7), by the National Academy of Science of Ukraine 'Problems of stabile development, rational nature management and environmental protection' (grant N 30-12), 'Fundamental problems of creation of new matters and materials for chemical industry' (grant N 21-13) and by the Ministry of Education and Science of Ukraine 'Development of technology of purification of liquid food and industrial waste waters with membrane methods' (grant N 262-14). The authors also thank Dr. N.N. Scherbatyuk (N.S. Kholodnii Institute of Botany of the NAS of Ukraine) for his help in research using TEM and SEM methods.

REFERENCES

1. Vibhandik AD, Marathe KD: Removal of Ni (II) ions from waste waters by micellar enhanced ultrafiltration using mixed surfactants.Front Chem Sci Eng 2014, 8:79.

2. Zeng J, Li S, Sun X, Chen X: Application of polyelectrolyte-enhanced ultrafiltration for rhenium recovery from aqueous solutions.Chem Eng Technol 2012, 35:387.

3. Hastuti E, Wardiha MW: A study of brackish water membrane with ultrafiltration pretreatment in Indonesia's coastal area.J Urban Environ Eng 2012, 6:10.

4. Bourgeous KN, Darby JL, Tchobanoglous G: Ultrafiltration of wastewater: effects of particles, mode of operation, and backwash effectiveness.Water Res 2001, 35:77.

5. Toker R, Karhan M, Tetik N, Turhan I, Oziyci HR: Effect of ultrafiltration and concentration processes on the physical and chemical composition of blood orange juice.J Food Process Preserv 2014, 38:1321.

6.   Baldasso C, Barros TC, Tessaro IC: Concentration and purification of whey proteins by ultrafiltration.Desalination 2011, 278:381.

7.   Kwon Y: Handbook of essential pharmacokinetics, pharmacodynamics and drug metabolism for industrial scientists. USA, Springer Science & Business Media; 2001.

8.   Mulder M: Basic principles of membrane technology. Kluwer, Dordrecht; 2000.

9.   Zeman LJ, Zydney AL: Microfiltration and ultrafiltration: principles and application. Marcel Dekker, New York; 1996.

10. Akbarnezhad S, Mousavi SM, Sarhadd R: Sol-gel synthesis of alumina-titania ceramic membrane: preparation and characterization.Indian J Sci Technol 2010, 3:1048.

11. Goosen MFA, Sablani SS, Al-Hinai H, Al-Obeidani S, Al-Belushi R, Jackson D: Fouling of reverse osmosis and ultrafiltration membranes: a critical review.Separ Sci Technol 2004, 39:2261.

12. Shi X, Tal G, Hankins NP, Gitis V: Fouling and cleaning of ultrafiltration membranes: a review.J Water Proc Eng 2014, 1:121.

13. Chen J, Ruan H, Wu L, Gao C: Preparation and characterization of PES-SiO_2 organic-inorganic composite ultrafiltration membrane for raw water pretreatment.Chem Eng J 2011, 168:1272.

14. Yu H, Zhang X, Zhang Y, Liu J, Zhang H: Development of a hydrophilic PES ultrafiltration membrane containing SiO_2-N-Halamine nanoparticles with both organic antifouling and antibacterial properties.Desalination 2013, 326:69.

15. Pang R, Li X, Li J, Zh L, Sun X, Wang L: Preparation and characterization of ZrO_2/PES hybrid ultrafiltration membrane with uniform ZrO_2 nanoparticles.Desalination 2014, 332:60.

16. Rahimi Z, Zinatizadeh AA, Zinadini S: Preparation and characterization of a high antibiofouling ultrafiltration PES membrane using OCMCS-Fe_3O_4 for application in MBR treating wastewater.J Appl Res Water Wastewater 2014, 1:13.

17. Kwak S-Y, Kim SH, Kim SS: Hybrid organic/inorganic reverse osmosis (RO) membrane for bactericidal anti-fouling. 1. Preparation and characterization of TiO2 nanoparticle self-assembled aromatic polyamide thin-film-composite (TFC) membrane.Environ Sci Technol 2001, 35:2388.

18. Dzyazko YS, Volfkovich YM, Sosenkin VE, Nikolskaya NF, Gomza YP: Composite inorganic membranes containing nanoparticles of hydrated zirconium dioxide for electrodialytic separation. Nanoscale Res Let 2014, 9:271.

19. Dzyazko YS, Belyakov VN, Stefanyak NV, Vasilyuk SL: Anion-exchange properties of composite ceramic membranes containing hydrated zirconium dioxide.Russ J Appl Chem 2006, 80:769.

20. Dzyazko YS, Mahmoud A, Lapicque F, Belyakov VN: Cr (VI) transport through ceramic ion-exchange membranes for treatment of industrial wastewaters.J Appl Electrochem 2007, 37:209.

21. Krzywonos M, Cibis E, Miśkiewicz T, Ryznar-Luty A: Utilization and biodegradation of starch stillage (distillery wastewater). Electronic J Biotechnol 2009, 12:1.

22. Acharya NK: Microscopic observation of nuclear track pores in polymeric membranes.Engineering 2011, 3:639.

23. Vilensky AI, Zagorski DL, Bystrov SA, Michailova SS, Gainutdinov RV, Nechaev AN:Investigation of latent tracks in polyethyleneterephthalate and their etching.Surface Sci 2002, 911:507-10.

24. Vilensky AI, Tolstikhina AL: Etching of tracks of accelerated heavy ions in poly(ethyleneterephthalate) and some physicochemical properties of track membranes.Rus Chem Bul 1999, 48:1100.

25. Dzyazko YS, Ponomaryova LN, Volfkovich YM, Sosenkin VE, Belyakov VN: Polymer ion-exchangers modified with zirconium hydrophosphate for removal of Cd^{2+} ions from diluted solutions. Separ Sci Technol 2013, 48:2140.

26. Dzyazko YS, Ponomareva LN, Vol'fkovich YM, Sosenkin VE, Belyakov VN: Conducting properties of a gel ionite modified with zirconium hydrophosphate nanoparticles.Russ J Electrochem 2013, 49:209.

27. Dzyazko YS, Ponomaryova LN, Rozhdestvenskaya LM, Vasilyuk SL, Belyakov VN:Electrodeionization of low-concentrated multicomponent Ni^{2+}-containing solutions using organic–inorganic ion-exchanger.Desalination 2014, 342:43.

28. Dzyazko YS, Ponomaryova LN, Volfkovich YM, Trachevskii VV, Palchik AV: Ion-exchange resin modified with aggregated nanoparticles of zirconium hydrophosphate. Morphology and functional properties.Micropor Mesopor Mater 2014, 198:55.

29. Ward AD, Trimble SW: Environmental hydrology. CTC Press LLC, Boca Raton; 2004.

30. Sata T: Ion exchange membranes. Preparation, characterization, modification and application. RSC, Cambridge; 2004.

31. Helfferich F: Ion exchange. Dover, New York; 1995.

32. Alberti G, Toracca F: Crystallyne insoluble salts of polybasic metals. 2. Synthesis of crystalline zirconium and titanium phosphates by direct precipitation.J Inorg Nucl Chem 1968, 30:317.

33. Gragg SJ, Sing KSW: Adsorption, surface area, and porosity. Academic Press Publisher, London, San-Diego; 1991.

34. Volfkovich YM: Influence of the electric double layer on the internal interface in an ion exchanger on its electrochemical and sorption properties.Soviet Electrochemistry 1984, 20:621.

Surface Analytical Approaches Contributing to Quality Assurance during Manufacture of Functional Interfaces

Kai Brune[1], Christian Tornow[1], Michael Noeske[1], Thorben Wiesner[1], André Felipe Queiroz Barbosa[1, 2], Stephani Stamboroski[1, 3], Stefan Dieckhoff[1], and Bernd Mayer[1]

[1]Adhesive Bonding Technology and Surfaces, Fraunhofer Institute for Manufacturing Technology and Advanced Materials IFAM, Wiener Straße 12, Bremen, D-28359, Germany

[2]Department of Mechanical Engineering, CEFET-MG, Federal Center for Technological Education of Minas Gerais, Av. Amazonas 7675, Nova Gameleira, Belo Horizonte, 30510000, MG, Brazil

[3]Department of Chemistry, UFSC, Federal University of Santa Catarina, Campus Universitário Trindade, Florianópolis, 8040-900, SC, Brazil

ABSTRACT

Assessing adhesion or strength of composites or adhesive joints in a non-destructive way is highly challenging. Therefore, instead of performing retrospective quality assurance, i.e. investigating manufactured joints, it is advantageous to safeguard performance and quality of each layer and each interface already during manufacture. This approach still is challenging, as it requires a systematic quantitative evaluation of threshold criteria, but appreciably it gets more and more feasible. We present approaches for an inline-capable and non-destructive quality assurance of steps in manufacturing processes used for tailoring the state of substrate surfaces. Benefits from applying techniques for inline surface analysis like Optically Stimulated Electron Emission (OSEE) and Aerosol Wetting Test (AWT) will be detailed. The performed procedures contribute to a novel class of non-destructive testing (NDT) techniques, classified as Extended NDT (ENDT). The principle of ENDT methods is based on the detection of selected physico-chemical properties which are important for the anticipated performance of the functional interfaces in the products to be manufactured.

A prerequisite for obtaining reliable composite materials is to reproducibly prepare a suitable surface state of the substrates before the first step of a coating or bonding process. As demonstrative application scenarios, we highlight first an exemplary surface pretreatment process for steel substrates, and second the identification of a surface state for carbon fiber reinforced polymer (CFRP) adherents suitable for joining. Concerning steel substrates we investigated two types of steel both in the as-received state and a state after grinding. We demonstrate that the removal of the topmost material layer comprising the reaction layer and mechanically deformed metal grains strongly affects the properties of the resulting adherent surface. As a consequence, a material-specific time slot for a steel substrate exposure in air after grinding is suggested in which the surface properties probed by OSEE remain unchanged. Moreover, we work out that the sensitivity and accuracy of inline-capable NDT techniques allow distinguishing surface states suitable for bonding of CFRP adherents from surface states unfavourable for adhesive bonding, and we exemplarily verify this statement for bonding processes applying freshly ground CFRP or, respectively, CFRP covered with thin layers of release agents.

BACKGROUND

Implementing reliable process steps like application, modification or removal of thin layers on substrates or adherents is crucial in a wide field of technological manufacture. Exemplarily, coating or adhesive bonding processes rely on a sequence of layer-oriented procedures comprising substrate cleaning, pre-treatment and, finally, the application of a fluid system which during the manufacture wets the substrate surface, hardens and forms a coating or an adhesive layer [1]-[4]. A huge variety of possible substrate materials is technologically used for distinct applications, ranging from steel for off-shore construction [5], bridge [6], pipeline [7] or ship building [8] to Carbon Fiber Reinforced Polymers (CFRP) for the manufacture of structural light-weight components in the aeronautics industry. The surface state of steel or CFRP substrates during coating or adhesive bonding governs the performance and quality of the manufactured products. Reliably coating steel substrates or joining modern polymer-based lightweight materials like CFRP using adhesive bonding technology will significantly profit from up-to-date quality assurance not only of the produced joints but also of the adherent surfaces entering the bonding process.

In the case of steel, the corrosion-protective effect of coatings on steel substrates strongly depends on the metal pre-treatment [9]. Concerning CFRP, in principle adhesive bonding is an optimum technique for joining light-weight structures based on heat-sensitive plastics, but difficulties in assessing the bond quality by non-destructive testing (NDT) limit the use for aircraft structural assembly. In consequence, certification by the regulation authorities is restrictive. For both steel and CFRP structures the coating or bonding process may be envisioned to take place starting from recently manufactured and thoroughly cleaned adherents in a well-climatised manufacturing site, or it may deal with already utilised devices when a repair is performed in a workshop environment.

Concerning steel, the surface state is checked for safeguarding the quality when manufacturing coated steel substrates or adhesive joints comprising steel adherents [13]. Investigations of steel surfaces using Optically Stimulated Electron Emission (OSEE) were reported during plastic deformation of steel samples in vacuum at low temperatures

[10], and a modified Geiger counter was used to measure the exoelectron emission from iron or nickel subjected to friction and wear or after exposure to gaseous environments like water vapor or oxygen at various pressures [11]. The OSEE signal was indicated to be sensitive not only to the state and thickness of the reaction layers on the metal surface [11] but also to contaminations, e.g. grease layers, on steel surfaces – in concentrations which result in effects on adhesion properties [12],[13]. Referring to parameters affecting the strength of metal/polymer interphases, for aluminum and titanium the initial bond strength of metal/polymer bonds and their long-term performance were reported to depend on the oxide layer formed on metal surfaces and its environmental stability, respectively [14], and for adhesive bonding with mild steel adherents it was shown that joints prepared under 7% r.h. result significantly stronger than joints prepared under 56% r.h. [15]. Especially when aiming at a pretreatment for bonding processes an increase of the surface roughness may be recommendable for improving the bond strength [16]. Thus, the surface preparation of steel substrates and their exposure before and during manufacture or repair will strongly influence the strength and durability of the resulting coated steel substrates or adhesive joints comprising steel adherents. Among mechanical surface preparation grit blasting, e.g. using abrasive impact of sand or corundum particles, or grinding are commonly used as pretreatment processes for steel adherent surfaces [17]-[20] or steel surfaces before painting [21]. In the present contribution, we apply a manual dry grinding process of cleaned specimens for the mechanically abrasive pretreatment of steel surfaces. We present the results of OSEE investigations of two distinct types of steel, and the as-received and a freshly ground surface state are characterized. Moreover, we line out the surface properties after distinct exposures to air and immersion in water, profiting from the advantage of the online technique that superficial changes can be detected very quickly. Related to the significance of reproducibly equal conditions to ensure the quality of bonding systems, we suggest a span for the open time before applying a coating or an adhesive system.

Concerning CFRP, manufacturing and in-service effects were shown to influence the mechanical performance of joints prepared from such polymer adherents [22],[23]. Exemplarily, within the ENCOMB project Markatos et al. demonstrated that the mode 1 (G_{1c}) interlaminar fracture toughness of adhesively bonded joints strongly depends on the surface state of the CFRP panels introduced into the bonding process. Exemplarily, when silicone-based release agents are used in a molding process during the manufacture of CFRP, silicone concentrations in the range of 5 to 20 atom % (according to XPS investigations) can typically be obtained [23]. Different further factors were shown to result in a decrease of the pristine $G_{1c,p}$ value. As compared to pristine dry and clean reference adherents, very pronounced effects were obtained when applying approximately 1 nm of a release agent, resulting in $G_{1c,p}$ being lowered by nearly 70%. These results are at the basis of the investigations detailed in this work, because before the adhesive bonding process the surface state of the adherents had been investigated applying NDT within an extended NDT approach. Therefore, the mode 1 (G_{1c}) interlaminar fracture toughness values reported by Markatos et al. are detailed in Figure 1.

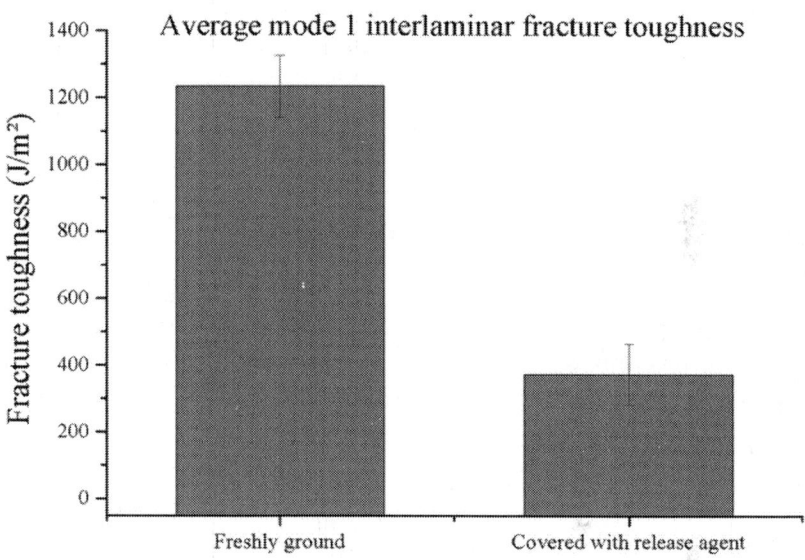

Figure 1: Mode 1 (G_{1c}) interlaminar fracture toughness values reported by Markatos et al.[23]for freshly ground CFRP adherents and CFRP adherents covered with release agent, respectively. Before the bonding was performed by these authors quality assurance measures reported in the present contribution had been performed.

Basically, such huge influence of contaminants on bond strength as demonstrated within the ENCOMB project clearly emphasizes the need to monitor the state of adherent surfaces before introducing them into a bonding process. A non-destructive approach may be offered by performing a water break test. However, the wetting behavior of the CFRP panels will be affected in a complex way by the roughness of ready-to-bond CFRP surfaces. Alternatively, Optically Stimulated Electron Emission (OSEE) permits a considerable performance for surface investigations of CFRP under ambient conditions, as detailed particularly by Parker and Waghorne [22]. However, OSEE has not yet achieved a widespread use for investigating the surface state of CFRP substrates exhibiting contaminations considered relevant, e.g., in aeronautical use. As a forward-looking contribution, the results presented here were obtained in the frame of the ENCOMB project and with CFRP panels prepared correspondingly and in parallel to the adherents used by Markatos et al. [23]. OSEE [24] and Aerosol-Wetting-Test (AWT) [25],[26] were assessed and advanced with respect to their applicability as ENDT techniques for sensitively indicating the presence of intentionally deposited and around 1 nm thin layers of a release agent on CFRP sample surfaces.

METHODS

In this chapter, experimental details are presented. The surface analytical tools are described, and the implementation of the scenarios for steel and CFRP applications is depicted.

Analytical Tools for Extended NDT f Surfaces

Optically Stimulated Electron Emission (OSEE) experiments were performed under ambient conditions with a Surface Quality Monitor SQM200 (purchased from Photo Emission Tech., Inc. (PET), USA). Concerning the principle of an OSEE measurement, the sample surface is exposed to ultra-violet light of a mercury vapour lamp with prominent emission maxima at 4.9 and 6.7 eV. Due to the work function of the respective substrate surface amounting to approximately 5 eV, the latter emission maximum essentially contributes to the photoelectrons emitted by the sample surface, and emitted electrons will exhibit kinetic energies of less than approximately 2 eV. Such photoelectron energy results in a clearly sub-micrometer information depth of this method when investigating the surface of a solid sample. The interaction of the emitted photoelectrons with the ambient atmosphere is predominated by an electric field effective at the sensor to an extent which permits sensor surface distances in the millimeter range during OSEE measurements. Carefully controlling the distance between the sensor and the surface is a prerequisite for effectively applying the set-up sketched in Figure 2.

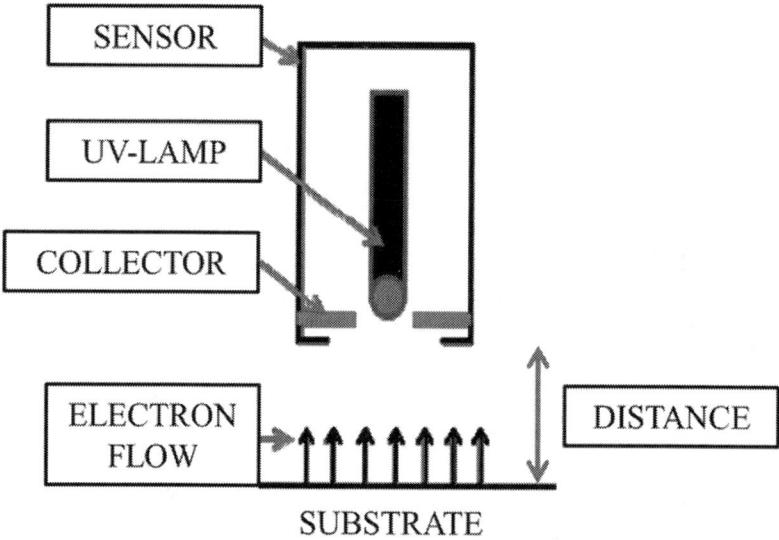

Figure 2: Sketch showing basics of a set-up relevant for performing non-destructive surface analysis using Optically Stimulated Electron Emission (OSEE).

For mounting the substrates to be analysed, the OSEE device is provided with an electrically conductive and earthed moving table, on which the analyte sample is positioned by movements in two perpendicular horizontal directions under the sensor. The vertical distance between the sample surface and the sensor is set using a micrometer screw attached to the holder of the sensor. Controlling the sample-to-sensor distance is essential during the OSEE measurements, and during the reported surface analytical investigations of the flat steel sheets the sample-to-sensor distance is maintained constant. A surface scan is then performed, with the table being programmed to move according to a step size and a number of steps (in both horizontal directions) defined by the user applying a software associated to the machine. As the scan advances, a photocurrent is obtained for each part of the scanned surface, and a dimensionless value further on denoted as the OSEE signal is indicated on a monitor. At the end, a digital worksheet with the emission values for the entire analyzed sample, i.e. an OSEE map, is presented as a result of the test performed. In detail, during the scanning of the steel samples the panels were positioned at a distance of 3.19 mm below the OSEE sensor, and when acquiring averaged OSEE intensities the parameters of the horizontal moving of the table were set as 1 mm step size and 25 steps in both directions.

The Aerosol Wetting Test (AWT) is a method to characterise the wetting properties of extended surfaces. The Aerosol Wetting Test was developed [25] to overcome drawbacks of contact angle measurements and the water break test. Depending on the surface state and surface energy, droplets of an aerosol form wide or narrow drops when applied to surfaces. For a defined liquid volume, the size of a single droplet depends directly on the contact angle. When several droplets are deposited, the property of a surface to be wetted can be characterised by the droplet size distribution. If the surface energy of a sample is low, e.g. due to contaminations like release agents, narrow drops will be formed on the surface. If the surface energy of a sample is high, wide drops will form. In the experiments, an ultrasonic atomiser nozzle was used to create an aerosol of small water drops with a narrow drop size distribution. This nozzle was fed by a syringe pump ensuring a constant water flow. By a constant airflow these drops are sprayed onto the surface of the sample. Depending on the surface energy of the sample, the area density of wide droplets increases with increasing the amount of water deposited. The resulting drops are imaged by a camera (Olympus, ColorView III) positioned perpendicular to the surface. Using an analysis software, the images are processed. The processing comprises background subtraction and quality refining; in this way structures or scratches on the sample are considered. Afterwards the diameters of the drop sizes are measured and classified. Using a two-parameter fit, the results are fitted to match a Rosin-Rammler drop-size distribution. The mean distribution width is used to rate the fitting results.

Additionally, steel surfaces and surfaces of CFRP panels were characterised using vacuum-bound instrumental analysis. Investigations of the surface composition were performed with X-ray Photoelectron Spectroscopy (XPS), and for XPS investigations pieces were cut out of the several centimetres wide original CFRP samples. XPS spectra with an information depth of around 10 nm were taken using a Kratos Ultra system applying excitation of photoelectrons by monochromatic $Al_{K\alpha}$ radiation within an area of approximately 0.2 mm². The system was operated at a base pressure 4×10^{-8} Pa, the sample neutralization was performed with low energy electrons (<5 eV). An electrostatic lens was used, the take-off angle of electrons was 0° with respect to the surface normal, and the pass energy was fixed to 20 eV (or, respectively, 40 eV in case of some less concentrated constituents) in high resolution spectra and 160 eV in survey spectra. Elemental ratios were calculated based on the area of the peaks and considering relative sensitivity factors.

Investigations of the surface structure of small CFRP and steel pieces were performed with Scanning Electron Microscopy (SEM), applying a Field Emission Scanning Electron Microscope (FESEM) of type FEI Helios 600 (DualBeam). The specified resolution is 0.9 nm at 15 kV at optimal working distance, and 1 nm at 15 kV at the coincidence point. Energy Dispersive X-ray (EDX) investigations with an information depth of a few micrometers were performed in 200 μm wide regions of steel samples using an acceleration voltage of 20 kV for the incident electron beam.

Implementation of Scenarios Relevant for Steel Application

Steel sample panels of type QD (ISO 3574 type CR1, CRS SAE 1008/1010) with a thickness of 0.5 mm and a smooth surface were obtained from Q-Lab Deutschland GmbH. According to the material specification the chemical composition of SAE 1008/1010 steel panels is based on iron containing maximum 0.60 weight% of manganese, maximum 0.15% of carbon, maximum 0.030% of phosphorus, and maximum 0.035% of sulfur [27]. As a second steel material stainless steel 1.4301 was purchased from Rocholl GmbH (Aglasterhausen, Germany), and its chemical composition is based on iron containing maximum 0.07 weight% of carbon, from 17.5% to 19.5% of chromium, and from 8.0% to 10.5% of nickel [28].

Cleaning of steel samples was done by wiping the surfaces with a cellulose tissue soaked with isopropanol. For the pretreatment of steel surfaces three kinds of grinding papers were used, in all cases waterproof silicon carbide papers with grit sizes of 80, 320 and 800 mesh, from Hermes Schleifmittel GmbH & Co. KG or from Struers GmbH respectively. The manual dry grinding process was subdivided in two steps. After evaluating the effect of grinding with papers exhibiting distinct grit sizes for the final finish, a two-step grinding process was established. The first step included the removal of the topmost material layer with a thickness of 10 to 15 µm – as determined gravimetrically - using the grinding paper with a grit size of 320 mesh. Subsequently, for obtaining a smooth surface finish the 800 mesh grinding paper was applied. Air exposure of ground steel surfaces was performed at room temperature at a relative humidity around 49% r.h.

Implementation of Scenarios Relevant for CFRP Application

Two scenarios of primary importance for aircraft manufacturers applying carbon-fibre reinforced polymers (CFRP) were investigated with respect to the requirements they impose for extended NDT technologies.

The CFRP material exhibited fibres arranged in UD layers and a thermoset matrix (T700 low density carbon fibers and HexPly® M21 matrix from Hexcel). Clean untreated reference CFRP samples were obtained by grinding until a fibre layer was reached and cleaned according to standards from aircraft manufacturers. Starting from such sample surfaces, as one of the factors effective in production layers or – after aspired cleaning - traces of silicone-based release agents were considered as a main scenario, and CFRP samples were dipped under defined conditions into a solution of Frekote 700 NC dissolved in hexane. The samples were dried for 30 min at room temperature and subsequently heated for 60 min at 80°C in an air circulating oven [23]. Based on XPS investigations, the coverage of CFRP surfaces with a well-defined amount of release agent was adjusted and confirmed at several surface positions.

RESULTS AND DISCUSSION

For the investigations detailed in this chapter, first Optically Stimulated Electron Emission (OSEE) was consciously selected to be the ENDT tool for investigating the surface state of steel samples, and the results of the respective investigations are lined out. Then, the capacity of two selected Extended Non-Destructive Testing (ENDT) devices for revealing the presence and amount of silicone-based release agent on Carbon Fiber Reinforced Polymer (CFRP) sample surfaces is highlighted, with the focus being on Aerosol Wetting Test (AWT) and OSEE investigations.

OSEE-Based Surface Quality Assurance for Steel Pre-Treatment Processes

In the frame of a preliminary survey, steel surfaces with distinct roughness were investigated by OSEE after dry grinding of the SAE 1008/1010 steel substrates with distinct SiC grit-based grinding papers. In detail, grit sizes of 80, 320 or 800 mesh, respectively, were applied in a one-step, dry grinding process which comprised both the removal of the topmost material layer and the surface finish. After the removal of approximately 20 µm of material, two sets of samples were prepared, and the samples of the first set were exposed to air for 1 minute whereas the samples of the second set were stored in air for 3 minutes. At least one minute of air exposure was applied because the scanning of the steel substrates during the obtained OSEE took around one minute. The obtained OSEE intensities are shown in Figure 3.

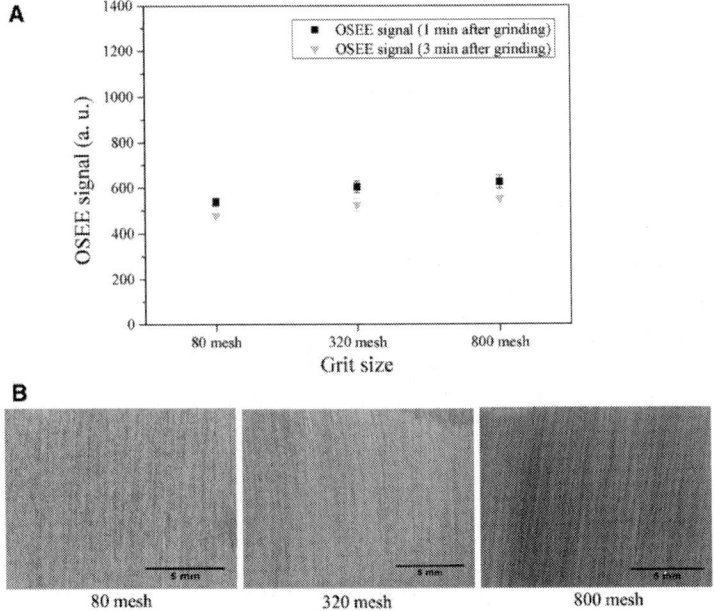

Figure 3: Results of non-destructive surface analysis using Optically Stimulated Electron Emission (A) and photographic documentation (B) for grinding surfaces of SAE 1008/1010 steel substrates using grinding papers with SiC grit sizes of 80, 320, and 800 mesh and then exposing them in air for 1 min or 3 min.

These results indicate that for all the applied grit sizes the OSEE intensity amounts to values around 580 a.u. after 1 min of air exposure and around 520 a.u. after 3 min of air exposure, with respective standard deviations between 20 a.u. and 40 a.u. of the OSEE intensity. From these findings we infer that – in contrast to the time period of air exposure - the roughness of the ground steel surface does not significantly influence the OSEE intensity.

The surface structure of SAE 1008/1010 steel samples was characterized, and in Figure 4 scanning electron microscopic results obtained are shown. The lateral contrast in the 100 μm wide SEM images is dominated by rather linear and parallel structures undulating, coarsely speaking, from the left side of the respective SEM image to the right side. These features are interpreted to result from the rolling process of the as-received steel surface and, respectively, the two-step grinding process in case of the surface finally ground with the 800 mesh SiC grinding paper. The rolling process appears to leave regions with some undercut of the mechanically flattened steel surface, and the grinding process here and there leaves blades with micrometer dimensions behind.

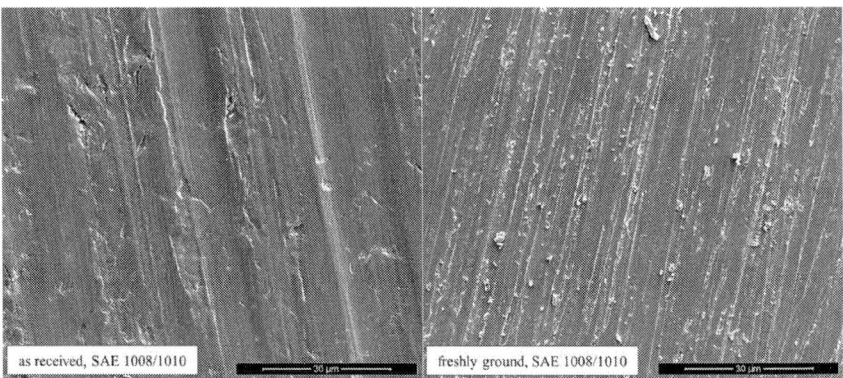

Figure 4: Scanning Electron Microscopy (SEM) based on secondary electron detection during imaging 100 μm wide regions of the as-received and the ground state of surfaces of SAE 1008/1010 steel.

In addition to changing the structure of the steel surface, the composition of the SAE steel surface after grinding differs from the composition before grinding. Following Table 1 and Table 2, evidence is furnished by EDX and supplemental XPS investigations which gather surface concentrations on the topmost surface layer with a thickness of a few micrometers and around ten nanometers, respectively. Generally speaking, in the as-received state manganese species are enriched close to the surface as compared to iron species, and grinding removes the surface region showing this enrichment and brings about an increase of the iron concentration close to the surface. XPS investigations indicate that after grinding the concentration ratio between metallic and oxidised iron species amounts to 0.44 which is approximately 15 times higher than in the as-received state. The surface termination of the as-received and cleaned steel panel appears to be characterised by a more than 8 nm thick oxide-based reaction layer.

Table 1: Concentrations of elements in a some micrometers thick surface region, given in atom% disregarding measured carbon contributions, as obtained applying Energy-Dispersive X-ray Analysis (EDX) investigations of SAE steel before/after grinding

Sample	Zn	Fe	Mn	Cr	Ca	Si	Cl	S	P	O	N	C	Ti	Cu
SAE steel, as-received and cleaned	-	95.7	0.7	0.2	-	0.2	-	-	-	3.0	-	--	0.1	-
SAE steel, ground, 72 h in air	-	96.8	0.4	0.1	-	0.7	-	-	-	1.7	-	--	0.1	0.2

Brune et al.

Brune et al. Applied Adhesion Science 2015 3:2, doi:10.1186/s40563-014-0030-0

Table 2: Concentrations of elements in an approximately 0.01 micrometer thin surface region, given in atom %, as obtained applying XPS

Sample	Zn	Fe	Mn	Cr	Ca	Si	Cl	S	P	O	N	C	$[Fe^0]/[Fe^\infty]$	$[Cr^0]/[Cr^\infty]$
SAE steel, as-received and cleaned	1.1	12.5	4.0	-	0.7	1.2	0.5	0.2	0.3	46.4	0.5	32.5	0.03	-
SAE steel, ground, 1 min in air	-	29.4	0.2	-	-	-	-	-	-	41.6	-	28.8	0.44	-
Steel 1.4301, as-received and cleaned	0.5	6.7	0.5	2.3	1.2	-	1.2	0.4	-	32.3	1.4	53.0	0.15	0.08
Steel 1.4301, ground, 1 min in air	-	20.5	2.2	7.9	-	-	-	-	-	37.3	1.2	30.9	0.41	0.30
Steel 1.4301, ground, 30 min in air	-	22.4	2.5	7.8	-	-	-	-	-	38.8	1.1	27.4	0.37	0.30
Steel 1.4301, ground, 4 h in air	-	18.1	1.7	6.7	-	-	-	-	-	41.2	0.6	31.7	0.28	0.25

The indicated values are due to averaging the results in two positions per sample, and signal ratios are given between metal-related and oxidized iron and chromium species.

Brune et al.

Brune et al. Applied Adhesion Science 2015 3:2, doi: 10.1186/s40563-014-0030-0

Moreover, Table 2 lists XPS results obtained when grinding steel samples of type 1.4301 and exposing them to dry air for distinct times. In the as-received state carbon-containing species dominate the topmost surface region. Iron and chromium are detected in a concentration ratio of approximately 3:1, and these constituents are detected both in the metallic as in oxidized states which indicates an average reaction layer thickness smaller than 10 nm and slightly thinner than in the case of the SAE steel. Grinding removes the surface region, and after grinding the concentration ratio between metallic and oxidised iron or chromium species is approximately threefold increased as compared to the as-received state. After one minute of exposure to air, the thickness of the reaction layer is similar as in case of the SAE steel, and after four hours the thickness of this oxide-based layer is significantly increased without having yet reached the reaction layer thickness of the as-received steel.

Effects of grinding the SAE steel and the 1.4301 steel samples, respectively, were also monitored using OSEE. While the OSEE map shown in Figure 5 indicates separate OSEE signals measured at individual positions of the sample surface, the signal intensities displayed in Figures 3, 6, 7 and 8refer to the average OSEE signal of the samples - obtained after a statistical analysis of the OSEE values measured at approximately 100 positions during scanning of the surface, and the standard deviation of this signal is displayed in the error bar. Moreover, the electron emission of the two types of steel was investigated under continued exposure in air and also in water.

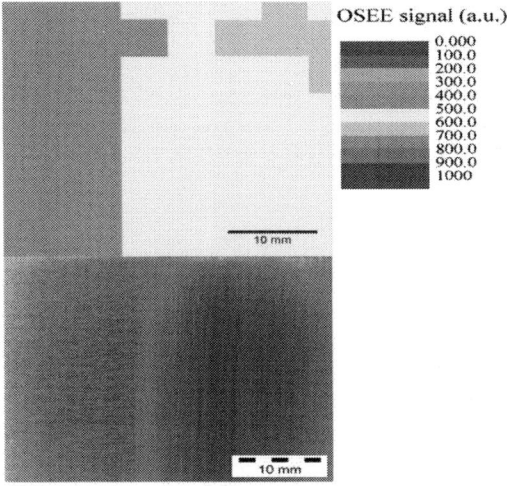

Figure 5: Results of non-destructive surface analysis using Optically Stimulated Electron Emission (OSEE) leading to an OSEE map (topmost image) displaying the local distribution of OSEE signals, and photographic image of the SAE 1008/1010 steel substrate surface. The surface states are as-received and cleaned (on the left side) and ground (on the right side).

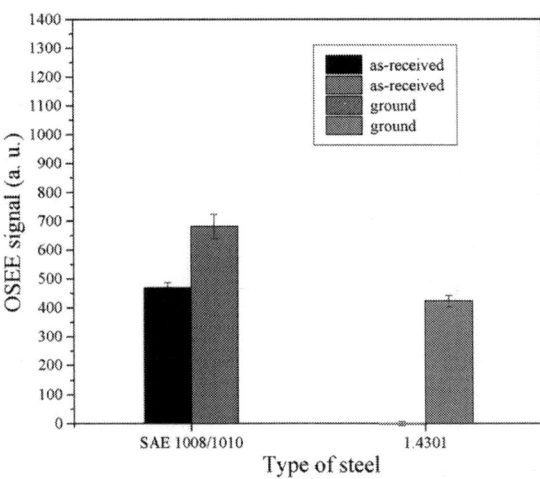

Figure 6: Results of non-destructive surface analysis using Optically Stimulated Electron Emission (OSEE) for two types of steel, namely SAE 1008/1010

steel and steel 1.4301, in two distinct surface states, namely as-received and ground.

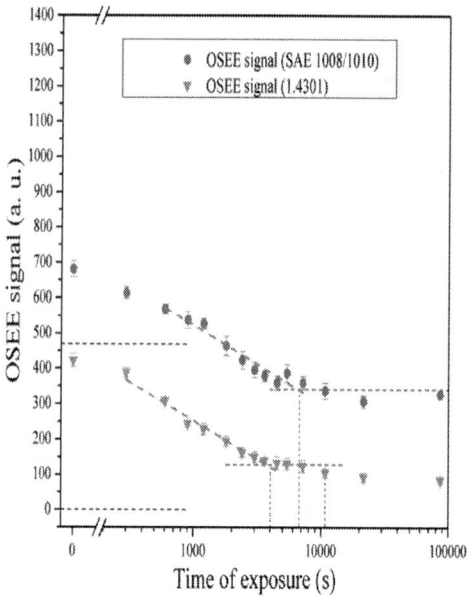

Figure 7: Results of non-destructive surface analysis using Optically Stimulated Electron Emission (OSEE) for freshly ground SAE 1008/1010 and 1.4301 steel during exposure in dry air. The horizontal dashed lines at the left side of the plot indicate the OSEE signals of as-received steel samples. Horizontal dashed lines at the right side indicate OSEE signals remaining basically unchanged after approximately two hours of exposure in air.

In the topmost part of Figure 5 an OSEE map displaying the local distribution of OSEE signals is shown, and below a photographic image of the investigated SAE 1008/1010 steel substrate is presented. The as-received and cleaned surface can clearly be distinguished from the ground surface state on the detail of an OSEE map measured with a step size of 5 mm of the moving table, as well as on the photograph.

Based on Figure 6 and the OSEE results displayed, we infer that OSEE investigations allow for differentiating two types of steel and their respective surface states since as-received or freshly ground SAE 1008/1010 steel shows different OSEE signals than as-received or freshly ground steel 1.4301. In detail, in both of the investigated surface states the steel SAE1008/1010 shows a higher signal than the steel 1.4301. And for both types of steel we observe a clearly increased electron emission as a consequence of the grinding process. These findings may be related to the SEM, EDX and XPS results presented in Figure 4 and in Tables 1 and 2, respectively. These investigations show that the grinding process results in strong changes of the structure and the composition of the substrate surface, and the XPS investigations indicate that exposure of freshly ground substrates of steel 1.4301 results in changes of the surface composition and, especially, in an increased thickness of the reaction layer on top of the surface. Consequently, we applied OSEE to characterize the surface state of freshly ground steel substrates during continued exposure in dry air. The obtained results are displayed in Figure 7 for the steel samples of type SAE1008/1010 and 1.4301, respectively. Coarsely speaking, the OSEE signal significantly decreases during at least one hour for both types of steel, according to a first regime of substrate behaviour. After this phase - in a second regime of substrate behaviour - for two hours no significant change of the OSEE signals is detected. A graphical evaluation determining the intersection of both of these regimes for both types of steels reveals that the OSEE signal of steel 1.4301 reaches the second regime significantly earlier than the OSEE signal of SAE1008/1010 steel. This observation indicates that the time span for reaching the intersection is material-dependent.

Finally, Figure 8 shows effects of water immersion on the OSEE signal obtained for freshly ground SAE1008/1010 steel. We infer that an immersion of the SAE steel panels in water clearly decreases the OSEE signal. As such effect on the surface state of the steel already is observed after 90 seconds of immersion in water, we decided not to apply a water-based AWT for a non-destructive testing of the steel samples. Moreover, as the changes of the surface state are effective much faster than during exposure of ground steel sheets in air we may infer that wet grinding processes will result in distinct surface states of steel substrates as compared to dry grinding.

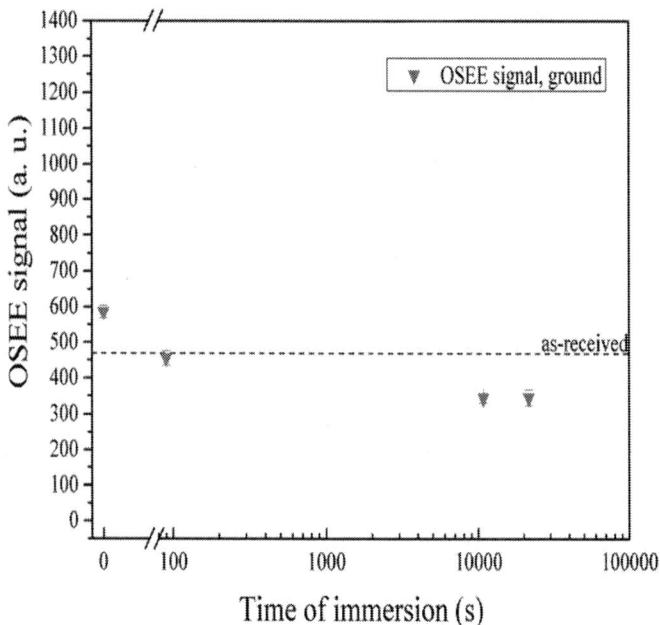

Figure 8: Results of non-destructive surface analysis using Optically Stimulated Electron Emission (OSEE) for freshly ground SAE 1008/1010 steel after immersion in water. The horizontal dashed line indicates the OSEE signal of an as-received steel sample.

Subsequently, we will discuss our findings with respect to their relevance for coating or adhesive bonding processes. Generally speaking, a reproducible state of the substrate surface may be obtained using as-received panels which had been stored in air for several weeks before being cleaned or panel surfaces obtained after grinding and exposure in air. As a mechanically abrasive pretreatment will remove surface layers on the as-received samples formed in long-term contact between the panels and the surrounding air, the freshly ground steel surface may need time to kind of get equilibrated in the atmosphere of the workplace used for the surface pretreatment. Therefore, the exposure in air may be considered a second step of a grinding process and needs to be inspected.

Considering that OSEE signals are influenced by changes of surface states, we used this technique to follow the influence of the exposure time in air after the grinding process. Such proceeding was intended to help in identifying a maximum and a minimum open time of freshly ground steel sheets exposed to air. Following Figure 7, during an increasing exposure time in air after the grinding of both types of steel the OSEE signal of the steel panel surfaces was followed. We observed a strong decrease of the OSEE signal after comparatively short times of exposure, and for longer times in air a steady state of the steel surfaces is reached resulting in a plateau of the OSEE signal.

This information is considered relevant if grinding is used as a pretreatment, bearing in mind that the reproducibility of the output of the coating or bonding processes starting from the ground steel adherents may depend on the open time after grinding. For interpretation from the chemical point of view, we may tentatively attribute a change in the OSEE signal after exposure in air or water to a change of the state of the substrate surface [11]. The X-ray Photoelectron Spectroscopy (XPS) investigations performed for the 1.4301 steel samples show that such change affects the elemental composition of the topmost atomic layers and the thickness of the reaction layer on top of the metal bulk. XPS investigations of SAE 1008/1010 steel will reveal if such observations hold true also for this type of substrate. In any case, from the engineering point of view the surface state of an adherent in a bonding process or a substrate in a coating process may be most reliably achieved if its change after the grinding is minimal. Thus, we might expect to obtain the most reliable joint properties if the time-dependent change of the OSEE signal is minimal, and the reported observations on changes of the OSEE signal are important for the surface quality assurance during the adhesive bonding or coating process, based on reproducibly and reliably achieving a steady surface state of the steel substrates. In detail, we infer that after grinding a steel substrate a material-dependent time span of exposure in air should be waited before applying the adhesive or coating system.

Extended NDT of Silicone-Based Release Agent on CFRP Surfaces

Concerning the starting state of CFRP surfaces, i.e. the untreated reference CFRP surface, before applying the selected scenarios, Figure 9 depicts that grinding resulted in carbon fibres being exposed at the surface, and area fractions each with approximately 50% were covered by carbon fibres and cured matrix resin, respectively.

freshly ground, CFRP   30 μm

Figure 9: Scanning Electron Microscopy (SEM) based on secondary electron detection during imaging 100 μm wide regions of the ground state of a carbon-fibre reinforced polymer (CFRP) surface.

Adjusting the coverage of a silicone-based release agent on CFRP sample surfaces sample preparation will be evidenced using X-ray Photoelectron Spectroscopy (XPS) which requires a destructive cutting of the CFRP samples before introducing them into an ultra-high vacuum chamber. Then the results obtained using recent NDT technologies for characterising the CFRP surfaces will be lined out.

Referring to the abscissa of Figure 10 and the respective error bars of the data points obtained, the successful application of well-defined amounts of the silicone-based release agent on ground CFRP substrates is demonstrated. While on the non-coated starting surfaces no silicon-containing species were detected, the XPS results reveal effective thicknesses of release agent in a range between approximately one nanometer and several nanometers.

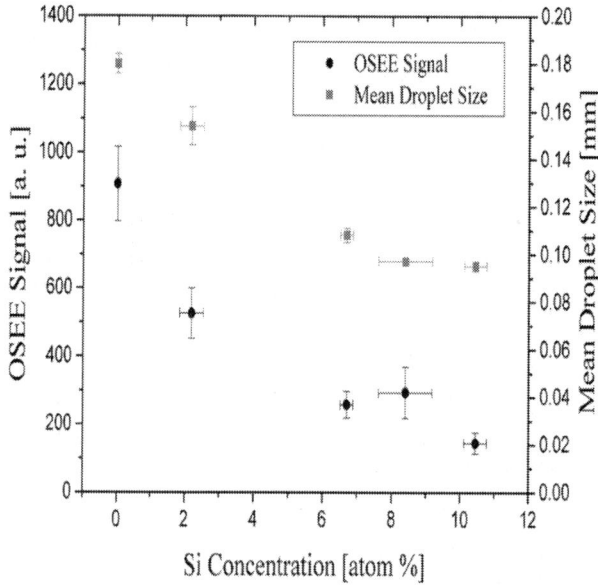

Figure 10: Results of surface quality assurance investigations of distinct states of carbon-fibre reinforced polymer (CFRP) surfaces using Optically Stimulated Electron Emission (OSEE) and Aerosol Wetting Test (AWT), respectively. The CFRP surface is covered with distinct amounts of a silicone-based release agent, as revealed by X-ray Photoelectron Spectroscopy (XPS) measurements.

Applying the NDT technologies Optically Stimulated Electron Emission (OSEE) and Aerosol-Wetting-Test (AWT) was achieved using the CFRP samples without further preparation before characterising their surfaces. Comparing the OSEE signal intensity with the concentration of silicon atoms as detected by XPS, the experimental data shown in Figure 10 clearly reveal that the signals obtained in OSEE studies are sensitive to the presence of even the thinnest layers of release agents investigated. Additionally, results from evaluating sizes of water droplets obtained from performing the AWT for CFRP surfaces free from release agent and contaminated with distinct amounts of release agent, respectively, are shown in Figure 10. For all the samples the same amount of water was applied, and the lateral dimensions and areas of imaged water drops show a size distribution. Basically, the portion of wider drops is observed to be significantly higher for the CFRP surface free from release agent which shows a better wettability with water as compared to the CFRP sample contaminated with release agent. The elaborate studies comprising a variation of the surface concentration of the release agent reveal that even at the lowest investigated thickness of the silicone-based films the average size of water drops is significantly smaller than in case of the CFRP surfaces free from release agent. That means that both OSEE and AWT investigations of CFRP adherents allow for clearly indicating the presence of a silicone-based release agent in an amount which – following Figure 1 – was demonstrated by Markatos et al. [23] to result in a strongly reduced strength of adhesive joints manufactured based on such adherents. These adherents therefore may be considered effectively contaminated. Cleaning them appropriately and testing the effect of the cleaning applying extended non-destructive testing will contribute to preventing the manufacture of adhesive joints with undesirably compromised strength.

CONCLUSIONS

Two recent non-destructive testing (NDT) technologies for the investigation of Carbon-Fibre Reinforced Polymer (CFRP) surfaces before adhesive bonding were assessed and advanced as a contribution to an extended non-destructive testing (ENDT) approach of composite bonds in the frame of the European FP7 project "ENCOMB – Extended non-destructive testing of composite bonds". Comparing to freshly ground CFRP surfaces, release agent contamination of the surface as a scenario relevant during CFRP application was considered. Both Optically Stimulated Electron Emission (OSEE) and Aerosol-Wetting-Test (AWT) showed a high potential as NDT tools detecting thin layers of a silicone-based release agent on Carbon-Fibre Reinforced Polymer (CFRP) surfaces. Investigations with either of these techniques applied to adherent surfaces before the adhesive bonding process allowed to indicate the inappropriate state of potential CFRP adherents, which according to Markatos et al.[23] was related to an application scenario reducing the joint strength of resulting adhesive joints.

In case of metal substrates prone to surface reactions in presence of liquid water, AWT investigations may not be appropriate to characterize the as-received surface state but only the surface state after wet chemical treatments. Concerning non-destructive testing of SAE1008/1010 and 1.4301 steel surfaces, the results of OSEE investigations were shown to depend on the type or composition of steel, to provide information about the application of a mechanically abrasive surface pretreatment like grinding, and to reveal changes of the surface composition during exposure of the freshly ground steel substrates in reactive environments like dry air or water. As an outlook, concerning the time period between performing a mechanically abrasive pretreatment of a metal substrate - especially steel substrate surfaces - and applying the adhesive to the metal surface we suggest monitoring the material-specific open time and documenting this parameter in operating procedures for bonding processes.

Concluding, for all these pre-bond processes engineers are facilitated to conduct appropriate revision of the adherent surface state applying the surface analytical technologies available in the frame of an extended NDT approach. Such proceeding contributes to ascertaining physico-chemical properties of adherent surfaces prior to bonding which are important for the performance of adhesive bonds. In this way, one concept aiming at quality assurance for adhesive bonding of composite or steel structures was demonstrated.

AUTHORS' CONTRIBUTIONS

KB coordinated the laboratory work for surface quality assurance, took care of the preparation of Carbon-Fibre Reinforced Polymer (CFRP) samples, and took part as well in the analysis and interpretation of the data as in drafting and revising the manuscript, CT performed Aerosol Wetting Test (AWT) measurements and the respective data evaluation, MN took part in defining and setting up the experiments with the steel samples, in analysing data and in drafting the manuscript, TW and AQ performed Optically Stimulated Electron Emission (OSEE) measurements and the respective data evaluation, TW and StSt performed the preparation and ageing tests of steel samples including their evaluation, StSt performed the lay-out and formatting of the article, TW contributed to data analysis and interpretation for the steel samples, SD and BM participated in the conception and design of the investigations and in providing information about technological boundary conditions of the surface pre-treatment and adhesive bonding processes. All authors read and approved the final manuscript.

ACKNOWLEDGEMENTS

The authors are grateful to Science without Borders (Ciência sem Fronteiras, A. F. Queiroz Barbosa 88888.049163/2013-00, and St. Stamboroski 88888.020610/2013-00) and to Coordination of Improvement of Higher Education Personnel (CAPES – Brazil). Research leading to the presented results received partial funding from the European Union's Seventh Framework Programme (FP7/2007-2013) under grant agreement number ACP0-GA-2010-266226 (ENCOMB, Extended Non-Destructive Testing of Composite Bonds). Finally, the authors acknowledge valuable hints given by a reviewer with respect to assessing steel substrates with distinct surface roughness.

REFERENCES

1. Brockmann W, Geiß PL, Klingen J, Schröder KB (2008) Adhesive bonding - materials, applications and technology. Wiley-VCH, Weinheim.

2. Markus S, Wilken R, Dieckhoff S, Hennemann O-D (2006) Quality monitoring of CFRP surfaces in bonding and coatings processes. In: Proceedings of the 16th international conference on surface treatments in the aeronautical and space industries, surfair, Bremen, Germany

3. Stenzel V, Rehfeld N (2011) Functional coatings. Vincentz Network, Hanover.

4. Schulz D (2010) Gut gereinigt ist halb geklebt. Adhäsion Kleben & Dichten 54(7–8):18-22

5. Momber AW, Plagemann P, Stenzel V, Schneider M (2009) Investigating corrosion protection of offshore wind towers. part 3: results of the laboratory investigations. J Protective Coatings Linings 26(11):38-47

6. Klinge R (2009) Altered specifications for the protection of Norwegian steel bridges and offshore structures against corrosion. Steel Construction 2(2):109-118 doi:10.1002/stco.200910015.

7. Yapp D, Blackman SA (2004) Recent developments in high productivity pipeline welding. J Braz Soc Mech Sci Eng 26(1):89-97

8. Weitzenböck JR, McGeorge D (2005) BONDSHIP project guidelines, 1st edn. Det Norske Veritas, Høvik

9. Elsner CI, Cavalcanti E, Ferraz O, Di Sarli AR (2003) Evaluation of the surface treatment effect on the anticorrosive performance of paint systems on steel. Prog Org Coat 48(1):50-62 doi:10.1016/s0300-9440(03)00112-7.

10. Sujak B, Olazowski Z, Dus-Sitek M (1983) Optically stimulated exoelectron emission and plastic properties of H17 steel. Radiat Prot Dosimetry 4(3/4):263-265

11. Momose Y, Namekawa T (1978) Exoelectron emission from metals subjected to friction and wear, and its relationship to the adsorption of oxygen, water vapor, and some other gases. J Phys Chem 82(13):1509-1515

12. Abedin MN, Welch CS, Yost WT (1991) Review of progress in quantitative nondestructive evaluation. Vol. 11B. In: Proceedings of the 18th Annual Review, Brunswick, ME, July 28-Aug. 2, 1991 (A93-19582 06-38), pp 1799–1805

13. Davis GD (1993) Contamination of surfaces: origin, detection and effect on adhesion. Surf Interface Anal 20:368-372

14. Venables JD (1984) Adhesion and durability of metal-polymer bonds. J Mater Sci 19:2431-2453

15. Gledhill RA, Kinloch AJ, Shaw SJ (1977) Effect of relative humidity on the wettability of steel surfaces. J Adh 9(1):81-85 doi:10.1080/00218467708075101

16. Baldan A (2004) Adhesively-bonded joints and repairs in metallic alloys, polymers and composite materials: adhesives, adhesion theories and surface pretreatment. J Mater Sci 39(1):1-49

17. Harris AF, Beevers A (1999) The effects of grit-blasting on surface properties for adhesion. Int J Adh Adhesives 19(6):445-452 doi:10.1016/s0143-7496(98)00061-X.

18. Poorna Chander K, Vashista M, Sabiruddin K, Sabiruddin K, Paul S, Bandyopadhyay PP (2009) Effects of grit blasting on surface properties of steel substrates. Mat Design 30(8):2895-2902 doi:10.1016/j.matdes.2009.01.014

19. Griffiths BJ, Gawne DT, Dong G (1996) The erosion of steel surfaces by grit-blasting as a preparation for plasma spraying. Wear 194(1–2):95-102 doi:10.1016/0043-1648(95)06798-1.

20. Multigner M, Ferreira-Barragáns S, Frutos E, Jaafar M, Ibáñez J, Marín P, Pérez-Prado MT, González-Doncel G, Asenjo A, González-Carrasco JL (2010) Superficial severe plastic deformation of 316 LVM stainless steel through grit blasting: effects on its microstructure and subsurface mechanical properties. Surf Coat Technol 205(7):1830-1837 doi:10.1016/j.surfcoat.2010.07.126.

21. Collazo A, Fernández D, Izquierdo M, Nóvoa XR, Pérez C (2005) Evaluation of red mud as surface treatment for carbon steel prior painting. Prog Org Coat 52(4):351-358 doi:10.1016/j.porgcoat.2004.06.008.

22. Parker BM, Waghorne RM (1991) Testing epoxy composite surfaces for bondability. Surf Interface Anal 17:471-476 doi:10.1002/sia.740170710.

23. Markatos DN, Tserpes KI, Rau E, Markus S, Ehrhart B, Pantelakis S (2013) The effects of manufacturing-induced and in-service related bonding quality reduction on the mode-I fracture toughness of composite bonded joints for aeronautical use. Composites Part B Eng 45(1):556-564 doi:10.1016.

24. Brune K, Lima L, Noeske M, Thiel K, Tornow C, Dieckhoff S, Hoffmann M, Stübing D (2013) Pre-bond quality assurance of CFRP surfaces using optically stimulated electron emission. Proceedings of the 3rd International Conference of Engineering Against Failure. ICEAF, Kos, Greece, p 300–307

25. Wilken R, Markus S, Amkreutz M, Tornow C, Seiler A, Dieckhoff S, Meyer U (2008) Method and device for testing a surface quality.

26. Garcia Gonçalves LM, Sanchez LC, Stamboroski S, Corrales Urena YR, Leite Cavalcanti W, Ihde J, Noeske M, Soltau M, Brune K (2014) Instantly investigating the adsorption of polymeric corrosion inhibitors on magnesium alloys by surface analysis under ambient conditions. J Surf Eng Mat Adv Technol 4:282–294

27. Q-Lab Corporation (2014). http://www.q-lab.com/de-de/products/q-panel-standard-substrates/q-panel-selector. Accessed 14 Nov 2014

28. Deutsche Edelstahlwerke GmbH (2014). http://www.dew-stahl.com/fileadmin/files/dew stahl.com/documents/Publikationen/Werkstoffdatenblaetter/RSH/1.4301_de.pdf. Accessed 14 Nov 2014

Citations

CHAPTER 1

Enrico Lertora, Chiara Mandolfino, and Carla Gambaro, "Mechanical Behaviour of Inconel 718 Thin-Walled Laser Welded Components for Aircraft Engines," International Journal of Aerospace Engineering, vol. 2014, Article ID 721680, 9 pages, 2014. doi:10.1155/2014/721680.

CHAPTER 2

Hui Lu, Dejun Ma, Jiasen Wang, and Jinghu Yu, "Research on Mechanical Behavior of Viscoelastic Food Material in the Mode of Compressed Chewing," Mathematical Problems in Engineering, vol. 2015, Article ID 581424, 6 pages, 2015. doi:10.1155/2015/581424.

CHAPTER 3

Firouzeh Sabri, Jeffrey G. Marchetta, K. M. Rifat Faysal, Andrew Brock, and Esra Roan, "Effect of Aerogel Particle Concentration on Mechanical Behavior of Impregnated RTV 655 Compound Material for Aerospace Applications," Advances in Materials Science and Engineering, vol. 2014, Article ID 716356, 10 pages, 2014. doi:10.1155/2014/716356.

CHAPTER 4

Md. Abu Mowazzem Hossain, Md. Tariqul Hasan, Sung-Tae Hong, Michael Miles, Hoon-Hwe Cho, and Heung Nam Han, "Mechanical Behaviors of Friction Stir Spot Welded Joints of Dissimilar Ferrous Alloys under Opening-Dominant Combined Loads,"Advances in Materials Science and Engineering, vol. 2014, Article ID 572970, 12 pages, 2014. doi:10.1155/2014/572970.

CHAPTER 5

B. Raju, B. Suresha, R. Swamy and B. Kanthraju, "Investigations on Mechanical and Tribological Behaviour of Particulate Filled Glass Fabric Reinforced Epoxy Composites," *Journal of Minerals and Materials Characterization and Engineering*, Vol. 1 No. 4, 2013, pp. 160-167. doi: 10.4236/jmmce.2013.14027.

CHAPTER 6

Xiaocong He and Yue Zhang, "Numerical Studies on Mechanical Behavior of Adhesive Joints," Advances in Materials Science and Engineering, Article ID 508135, in press.

CHAPTER 7

Jean C Batista Abreu, Luiz M C Vieira, Metwally H Abu-Hamd, and Benjamin W Schafer, Review: Development of Performance-Based Fire Design for Cold-Formed Steel, doi:10.1186/s40038-014-0001-3.

CHAPTER 8

Yuliya S Dzyazko, Ludmila M Rozhdestvenskaya, Yu G Zmievskii, Alexander I Vilenskii, Valerii G Myronchuk, Ludmila V Kornienko, Sergey V Vasilyuk, and Nikolay N Tsyba, Organic-Inorganic Materials Containing Nanoparticles of Zirconium Hydrophosphate for Baromembrane Separation, doi:10.1186/s11671-015-0758-x.

CHAPTER 9

Kai Brune, Christian Tornow, Michael Noeske, Thorben Wiesner, André Felipe Queiroz Barbosa, Stephani Stamboroski, Stefan Dieckhoff, and Bernd Mayer, Surface Analytical Approaches Contributing to Quality Assurance During Manufacture of Functional Interfaces, doi:10.1186/s40563-014-0030-0.

Index